THE EYE OF WAR

The Eye of War

Military Perception from the Telescope to the Drone

ANTOINE BOUSQUET

University of Minnesota Press
Minneapolis
London

A different version of chapter 2 was published as "Lethal Visions: The Eye as Function of the Weapon," *Critical Studies on Security* 5, no. 1 (2017): 62–80; copyright York University, Canada; reprinted by permission of Taylor & Francis Ltd, www.tandfonline.com on behalf of York University, Canada.

Published by the University of Minnesota Press
111 Third Avenue South, Suite 290
Minneapolis, MN 55401-2520
http://www.upress.umn.edu

Printed in the United States of America on acid-free paper

25 24 23 22 21 20 19 18   10 9 8 7 6 5 4 3 2 1

Library of Congress Cataloging-in-Publication Data
Names: Bousquet, Antoine J., author.
Title: The eye of war : military perception from the telescope to the drone / Antoine Bousquet.
Description: Minneapolis : University of Minnesota Press, [2018] | Includes index.
Identifiers: LCCN 2018008935 (print) | ISBN 978-1-5179-0346-6 (hc) | ISBN 978-1-5179-0347-3 (pb)
Subjects: LCSH: Military engineering—History. | Fire control (Gunnery)—Optical equipment. | Photography, Military. | Military topography. | Targeted killing.
Classification: LCC UG15 .B68 2018 (print) | DDC 355.4/13–dc23
LC record available at https://lccn.loc.gov/2018008935

You see, control can never be a means to any practical end . . . It can never be a means to anything but more control . . . like junk . . .

—WILLIAM S. BURROUGHS, *Naked Lunch*

CONTENTS

INTRODUCTION

Visibility Equals Death

The succinct outline of a curious military device can be found among the yellowing pages of an 1870 article published in the scientific journal *Nature.* Eleven years before, the Austrian army had reportedly set up a so-called electric torpedo defense to protect Venice from naval attack during the Franco–Austrian War. Rather than the self-propelled weapon it brings to mind today, the term "torpedo" refers here to the underwater naval mine, then still a largely experimental technology. The true inventiveness of the defensive system lay, however, in its particular use of a camera obscura, a darkened room in which a single small-lens aperture allowed for the focused passage of light to project an image of the outside world on a carefully positioned table (Figure 1). The article recounts that the

> camera obscura was erected in proximity to the harbour in such a manner that the horizontal table of the instrument reflected the whole area of the channel. Large wooden cases, each containing 400 pounds of gun-cotton, were lowered at certain fixed distances into the water, and as these disappeared one by one, a small row-boat described at the time a circle round the spot to indicate the extreme confines of the distance at which the torpedo would prove effective; an observer was stationed in the camera as these operations were going on, carefully watching their reflection in the instrument, and as each torpedo disappeared into the water, he marked with a pencil its precise locality on the white table, tracing also the ring formed by the row-boat. Thus a series of circles was formed in the camera, each of which was marked with a distinctive number, and in this way a miniature, but exceedingly correct, plan of the obstructions in the harbour was prepared; the wires in connection with the torpedoes were afterwards led up into the camera obscura and furnished with

> numbers to correspond with the circles. By means of this arrangement a sentinel stationed in the apparatus might at once explode any one of the torpedoes, as soon as he observed the reflection of an enemy's ship pass within the limits of the circles marked upon the table. The channel itself was quite clear of any suspicious buoys and beacons, and appeared to the enemy wholly free from obstruction.[1]

We do not know whether this arrangement actually functioned as intended, as it apparently was never tested in battle or replicated elsewhere. Yet its comparatively simple design enacted an ingenious and revealing articulation of perception, visual representation, and geolocation to assemble a covert and potentially lethal apparatus. Encapsulated within it is the programmatic ideal of an enduring martial dream: to see without being seen, to instantly apply deadly force at a remove, to conjugate vision with annihilation.

A century and a half later, this dream has essentially become a reality, with only its fullest planetary reach yet to be attained. At the outset of the new millennium, the United States Department of Defense announced that it was seeking a "prompt global strike" capability for the delivery of conventional munitions to any fixed or mobile target on earth within an hour of launch and to an accuracy of just a few meters.[2] Intended to dispense with forward-based forces, the program has elicited proposals for a range of delivery systems, from repurposed nuclear ballistic missiles to future hypersonic glide vehicles and even prospective space-based kinetic weapons. But regardless of the armament to be used, a precise and responsive strike capability across the planet's expanse would be inconceivable without a dense array of perceptual technologies permitting the timely detection, identification, location, and tracking of its targets.

The contemporary battlespace, a domain increasingly coterminous with the globe, is submitted to the intense survey of a *martial gaze* that threatens anything that falls under it with obliteration. Speaking in front of the Senate Committee on Armed Services in 1978, the undersecretary of defense, William Perry, declared that the Pentagon was pursuing the capability "to be able to see all high-value targets on the battlefield at any time, to be able to make a direct hit on any target we can see, and to be able to destroy any target we can hit."[3] While total transparency of the battlespace remains stubbornly elusive, a conver-

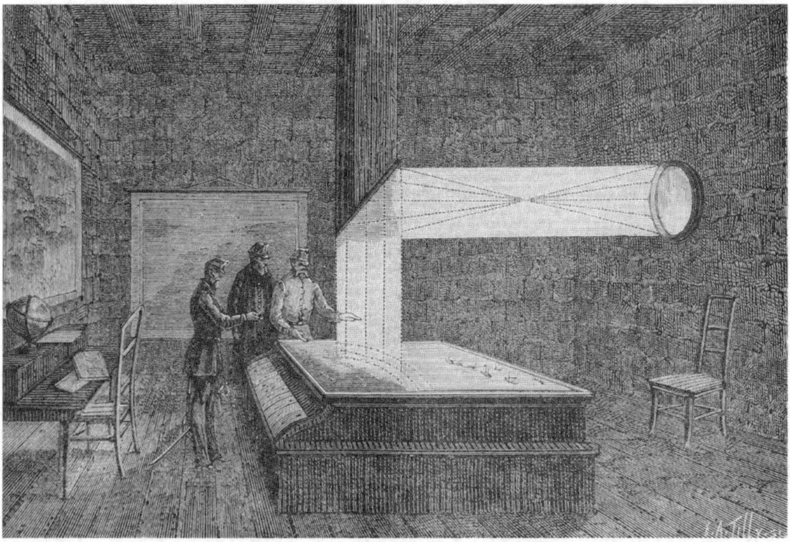

FIGURE 1. *Artist's impression of the Venetian electric torpedo defense. "The Camera Obscura in Torpedo Work,"* Nature *31, no. 800 (February 26, 1885). Image from the Biodiversity Heritage Library. Digitized by Smithsonian Libraries. www.biodiversitylibrary.org.*

gence of perception and destruction has in effect been realized today. Indeed, this convergence occupies a central role in the practice of global targeting, an emergent modality of military violence that is characterized not only by the planetary scale of its operations but also by a granularity that scales all the way down to the bodies of single individuals.

How have we come to inhabit a martial universe where, in Martin Libicki's pithy formula, "visibility equals death"?[4] What are the imperatives, knowledges, and mechanisms that compose the sociotechnical history of our present condition? How is the activity of soldiering transformed by the advent of a technological sensorium that both augments and envelops human perception? How are human bodies and minds mobilized, regimented, and redeployed as they become ever more tightly interwoven with their machines of war? Finally, what are the ultimate implications for global conflict and security of the advent of ubiquitous surveillance and targeting? What becomes of war as it increasingly comes to hinge on struggles over visibility across planetary battlespaces?

Global Targeting and the Disappearance of the Battlefield

The use of unmanned aerial vehicles (UAVs), as part of the United States' sustained campaign of targeted killings, has drawn substantial scholarly attention to the growing role of "lethal surveillance" in war.[5] In particular, an abundant academic literature on so-called drones highlights the articulation of remote vision and deadly force that characterizes the armed variants of such systems.[6] This alignment of perception and lethality is hardly novel in itself, however. Today's hunter-killer drone constitutes just its latest high-profile materialization. Indeed, it is endowed with features that are all tributary to preexisting technologies. Nonetheless, as long as one avoids fetishizing the drone as a somehow unique or causally determinative technology, it remains an illuminating object of analysis for current trends in the present-day conduct of warfare.

Considerable analysis has grappled with how to understand a weapon system deployed across the planet and able to strike at targets located outside of recognized war zones. Sensing a shift in the practices and norms of military force no longer legible within the traditional matrix of war, scholars have variously spoken of an "everywhere war,"[7] a "predator empire" of global surveillance and control,[8] a "cynegetic war" of planetary manhunt,[9] and the constitution of a "global battlefield."[10] All these notions can be regarded as various expressions of the battlefield's advanced dissolution as a discrete physical and normative space insulating military violence from the rest of society.[11] Put simply, as the practice of targeting becomes more globalized and individualized, the notion of distinct temporal and spatial bounds for the exercise of armed force becomes ever more untenable. The manifest reconfiguration of war denoted by this tendency is unquestionably a multifaceted phenomenon with a complex set of geopolitical, legal, strategic, and technological dimensions that should not be reduced to any one of them.[12] Yet the ascendancy of the martial gaze is indisputably one of its essential facets.

Indeed, perceptual technologies, in permitting the remote identification and location of discrete entities, have decisively contributed toward the individualization of targeting prosecuted on a planetary scale.[13] As Grégoire Chamayou explains, the traditional battlefield, conceived as a broadly delineated space in which armed combatants clash, has given way to ever more granular and time-sensitive "kill boxes" within which deadly force can be unleashed at will for the brief interval that they remain open.[14] On its initial operational deployment during the 1991 Persian Gulf War, the kill box was a vast expanse of around 900 cubic kilometers (roughly the area of New York City), patrolled for the aerial interdiction of the adversary state's armed forces.[15] At the turn of the millennium, the concept evolved to refer to a "three-dimensional area reference that enables timely, effective coordination and control and facilitates rapid attacks."[16] This more flexible and spatially restricted use of the kill box was notably deployed in the aerial support of the initial invasion of Iraq in 2003. With the drone's capability for lethal surveillance, the kill box's field of application has now migrated from the exclusive domain of designated war zones to, at least in principle, any location that presents a target of opportunity. Concurrently, the kill box's scale has been contracting down to that of a target individual's body, an exercise that the American military refers to as putting "warheads on foreheads."[17]

The emerging capacity to "credibly threaten and effectively conduct global strike by holding any target on the planet at risk" has far-reaching tactical and strategic ramifications.[18] In particular, the latent menace of ubiquitous surveillance and lethal targeting may dissuade an adversary from a course of action that would draw attention to its presence. As Paul Virilio muses, "If aiming is more important than the weapon itself, it is understood that one day deterrence will no longer be caused by weapons but by the gaze."[19] Yet as much as the wielders of the martial gaze might hope for its deterrent effect to be absolute, the vulnerability caused by visibility is instead merely shifting the terrain of struggle. While the conventional battlefield may be dissolving, there is no sign that conflict itself will follow.

The U.S. military's obliteration of the Iraqi army during the Persian Gulf War was undoubtedly an object lesson in the futility of fighting a mid-twentieth-century industrial war against an opponent at the forefront of the technologies of military perception. Indeed, none of the United States' adversaries has since been foolhardy enough to do so again. Instead, they have sought to offset the American superiority in the perceptual domain through various means. Nothing about this should surprise us, although we may simultaneously be troubled by the ramifications of these responses. Developments in the reach of military perception and its associated weapons have always been accompanied by the means to counter them. While these include a plethora of technical measures in the areas of camouflage and decoying, they also consist of more fundamental strategic responses for negating the transparency of the battlefield. Thus we observe growing practices of generalized concealment and dispersal, which may involve the covert infiltration of an adversary's civil society in advance of its unannounced assault. Global targeting is increasingly twinned with a scattering and burrowing of the enemy into the broader community of noncombatants that the worldwide projection of force purports to be defending. In tandem, these practices erode the spatially contiguous and temporally bracketed experience of conventional armed conflict. A new indefinite condition is emerging in its place, marked by discontinuous and punctuated eruptions of violence across a world in which the categories of war and peace, domestic and international, military and civilian are becoming ever more indeterminate. As a task force on U.S. drone policy led by the former head of United States Central Command conceded in 2014, "basic categories such as 'battlefield,' 'combatant' and 'hostilities' no longer have clear or stable meaning."[20]

The Logistics of Military Perception

In his seminal work *War and Cinema,* first published in 1984, Paul Virilio memorably formulated the notion of a "logistics of perception" that he identified as central to the development of modern warfare.[21] As he explained in a later interview, "The idea of logistics is not only about oil, about ammunitions and supplies but also about images. Troops must be fed with ammunitions and so on but also with information, with images, with visual intelligence. Without these elements troops cannot perform their duties properly. This is what is meant by the logistics of perception."[22] Although often neglected by popular accounts of war, logistics has long constituted a vital component of military operations in organizing the transport, supply, and maintenance of armed forces, a role that has only grown with war's increasing mechanization and technical sophistication.[23] The unprecedented endeavors in this area prompted by World War II directly gave rise to the modern business science of logistics that today supports globalized economic production and commercial trade.[24] For Virilio, war's evolution over the past century has made logistics a paramount activity, imposing its logic, needs, and modes of operability. The mobilization of industrial and popular resources determined the outcome of the two world wars, with victory awarded to the side that could bear punishing attrition the longest. During the subsequent Cold War, a fragile nuclear stalemate demanded of each side exhaustive preparation and perpetual readiness to unleash the full destructive might of their arsenal. Virilio concludes that "if we can say that war was entirely strategy in past societies . . . we can now say that strategy is no more than logistics" and that "logistics has become the whole of war."[25]

Whether we take this sweeping statement at face value, logistics' growing centrality in war is unquestionable. The writings of nineteenth-century military theorist Antoine-Henri de Jomini contain the first systematic treatment of logistics understood as "the practical art of moving armies."[26] While such a definition, with its exclusive focus on campaigning, might be considered narrow by contemporary standards, Jomini crucially attributed the essential character of a science to logistics, beginning with the calculation of distances and durations involved in marches. An earlier nonmilitary usage of the term already referred to the "art of arithmetical calculation," drawing its etymology from the Greek *logistikós,* "skilled in calculating."[27] Thus the will to scientific ordering and rational optimization is one of logistics' fundamental traits.[28] Accordingly, Nigel Thrift offers us a useful modern definition of logistics as "a set of knowledges synonymous with movement, effectively the science of moving objects in an optimal fashion."[29] Taken seriously, such a conception upends the general presumption that military logistics ends where combat begins—or, in professional parlance, where the tail of the military leviathan gives way to its teeth. An enlarged definition of logistics might then encapsulate not only the timely supply of matériel to a given location but also the opportune delivery of targeted force to discrete positions in space.

At any rate, the understanding of logistics must certainly include the orderly movement of information and images. When Jomini first commented on the growth of logistical activity occasioned by the increased mobility and spatial extension of armies, he underlined "the duty of transmitting the conceptions of the general to the most distant points of the theater of war and of procuring for him the necessary plans of operations."[30] Indeed, no effective movement of troops and matériel is possible without reliable procedures for gathering intelligence and distributing orders: in effect the systematic collection, storage, and transmission of information. While the technological means for such activities were limited in Jomini's time, the processes and techniques available for the production, representation, and circulation of information and sense perception have since exploded. The "logistics of perception" is therefore not a mere metaphorical turn of phrase but a quite literal description of the perceptual field's organization in modern warfare.

This book seeks to uncover and detail the historical constitution and present operation of this logistics of perception and its attendant exercises in targeting. Three key functional elements work together to align perception and destruction: sensing, imaging, and mapping. Sensing refers to the processes of sense perception by which an environment and the entities within it can be discerned. It includes the apprehension of sensorial stimuli channeled through visible light, the wider electromagnetic spectrum, sound waves, or any other detectable medium. Imaging denotes the means by which the information obtained through sense perception is recorded, ordered, and represented, thereby enabling its subsequent storage, dissemination, and processing. Finally, mapping encompasses the activities that correlate sense perception and its representations to the relative locations of objects and places within physical space. Materially instantiated in a multiplicity of different sociotechnical systems, the articulation of these three functions underlies the logistics of perception in the military targeting of force.[31]

While formally distinguishable, sensing, imaging, and mapping are in practice deeply intertwined activities in terms of both the complementarity of their operations and the entanglement of their genealogies. Indeed, all three functions already combine seamlessly within ordinary human vision, where the ocular lens focuses light to form an image on the eye's retina and the brain spontaneously estimates proportions and distances within the viewed scene. If this book's organization treats the functions of sensing, imaging, and mapping sequentially, their crucial mutual dependencies will cause their respective accounts to routinely blend into one another, all the more as the informational flows that join them tighten.

Similar considerations apply to the final dimension of the logistics of perception, the countermeasures devised to evade the martial gaze. Although such subterfuges do not fall directly within the core functions outlined above, they constitute an essential spur to the martial gaze through an intense dialectical relationship. War is an activity that emerges from a struggle of opposing forces. Any new source of combat power inevitably stimulates efforts to undercut or negate it. Significant expansions of the perceptual field have therefore always called forth inventive responses designed to thwart and offset any military advantage. The plethora of schemes conceived for concealment, masquerade, and decoying are all forms of hiding. Together, these undertakings constitute the obverse to the martial gaze. They draw on the same bodies of knowledge and technique as the gaze, but they exploit their existing biases and shortcomings for the purpose of concealment. Military perception evolves in turn by the necessity of foiling these countermeasures. The fundamental dynamic underlying the martial gaze's mutations is thus of a perceptual arms race pitting hiders and seekers against each other. At stake in this spiraling contest lies the fate of the human itself.

Man, the Machine, and War

Throughout the modern era, the perceptual nets cast over the battlespace have gradually apprehended a broader gamut of sensory stimuli, most notably through the advancing mastery of the electromagnetic spectrum's frequency range beyond that of visible light. They have steadily extended the physical distances for capturing such stimuli, as exemplified by contemporary remote-sensing satellites' minute observations of the terrestrial surface. Concurrently, innovations in perceptual technology have made possible increasingly fine discriminations within the span of detectable stimuli, allowing for the identification of ever fainter target signatures. These cumulative developments result, in the first instance, from advances in the scientific understanding of the physical world and various feats of engineering prowess. Underlying all of them is a rational organization of the perceptual field that has reduced sensorial activities to certain logical rules and procedures amenable to mechanization and automation. This rationalization of perception began with the visual faculty that Leonardo da Vinci believed to be the highest of the senses and the royal road to human knowledge and mastery of the world[32]—or as David Levin puts it, "The will to power is very strong in vision. There is a very strong tendency in vision to grasp and fixate, to reify and totalize: a tendency to dominate, secure and control."[33]

A substantial literature has long highlighted and critiqued vision's privileged role, or ocularcentrism, in the constitution and self-understanding of modern societies.[34] In particular, much has been made of the notion of a dominant "scopic regime" characterized by a "Cartesian perspectivalism, with its reduction of perceptual space to mathematical and homogeneous space, with its understanding of vision as monocular, static, fixed and immediate, distant and objectifying, purely theoretic and disincarnated."[35] Conjoining the rigorously geometric art of the Renaissance and the burgeoning scientific worldview of the early modern period, this rationalized articulation of vision and space resonates strongly with the analysis advanced here.[36]

A number of scholars have more recently appropriated the concept of the scopic regime to illuminate the contemporary exercise of armed force. They have variously defined the scopic regime as the "mode of visual apprehension" of "a militarized regime of hypervisibility,"[37] the "visual practice" that "operationalizes the kill chain,"[38] and the "grid of intelligibility" through which militaries "en-vision the battlespace."[39] The terminology of the "scopic" is certainly apropos to the subject matter. The etymology of the word can be traced back to the Greek *skopos,* which refers to both a "watcher"—notably in the military roles of the "scout" or "spy"—and a "target," "the mark or object on which one fixes the eye."[40]

Principally spurred by the high-profile use of drone aircraft, this contemporary visual turn in the study of war via scopic regimes is a welcome development. Without fundamentally disagreeing with the aforementioned scholarship's valuable contributions, this book will privilege the term "martial gaze" for three main reasons. First, it underlines the fact that the history of the rationalization, mechanization, and eventual autonomization of perception has always been entangled with—and from the twentieth century on decisively accelerated by—military imperatives. Second, while previous studies in cultural theory have emphasized the social, cultural, and even aesthetic dimensions of militarized vision in congruence with the origins of scopic regimes, the present book's primary focus is firmly on the technoscientific foundation and functional operation of the different constituents of the logistics of perception. This is not to discount the wider political and sociocultural contexts to these developments; indeed, they are evoked at various points. They are simply afforded less attention here than the immanent logic underlying the convergence of perception and destruction and the sociotechnical field of possibility traced by this conjunction. This book's ultimate wager is that by plunging into the heart of the machine, we may obtain a truer sense of the potential and limits of our agency within it, political or otherwise. Finally, while the visual domain occupies a preeminent place in its genealogy, military perception cannot be reduced to it, which a reliance on scopic regimes might lend itself to. The concept of the gaze has always had a more metaphorical resonance in its various theoretical usages, even if its deployment in this book is not directly tributary to any of them.[41] The wording of "martial gaze" should therefore not be taken literally as denoting an exclusive focus on the visual; rather, it instead encompasses the entire range of sensorial capabilities relevant to the conduct of war. Certainly it could be rejoined that the representations of this extended sensorium are still overwhelmingly visual, but this would miss the crucial fact that the operation of the gaze increasingly circumvents the human subject and "sees" without any need for visual mediations.

Indeed, one of the book's overarching narratives is perception's progressive relocation from an exclusive residence within its native biological substrate to myriad technical apparatuses. The human senses, which originally inspired the search for their artificial replication, are increasingly demoted or even altogether displaced by devices to which they are, often unfavorably, compared. In a public lecture delivered in 1868, prominent German scientist Hermann von Helmholtz, drawing on an analogy dating back to the sixteenth century, noted that "as an optical instrument, the eye is a camera obscura."[42] From his extensive ophthalmological studies, he concluded that the eye was a highly imperfect implementation of its underlying optical principles. "If an optician wanted to sell me an instrument which had all these defects," he opined, "I should think myself quite justified in blaming his carelessness in the strongest terms, and giving him back his instrument."[43] Helmholtz did go on to insist that these shortcomings were largely compensated by the eye's dynamic movement and the unconscious mental apprehension of its images, granting human vision a considerably greater acuity than suggested by the sole design of its organ. Nevertheless, his work participated in the institution of a systematic scientific evaluation of perceptual performance, establishing the eye as just one sensorial instrument among others. As these human faculties are found to be deficient or inferior relative to their artificial counterparts, so are they ineluctably supplemented or wholly replaced in the execution of specific tasks.

In the martial context, the human sensorium has been slowly and surely directed, mediated, and supplanted in service to the ultimate imperative of targeting. At one end, automated weapons such as the self-guided missile are now equipped with their own perceptual organs that seek, identify, and pursue their targets independently. For Virilio, such devices constitute a "self-sufficient gaze" with "the machine looking for itself and no longer for some spectator or telespectator."[44] At the other end, the human eye is disciplined and regimented through a process that John Pettegrew refers to as "the transformation of seeing as primary mode of human perception to seeing as military targeting."[45] Prosthetically tethered to the war machine, the combatant's cognitive and neurological labors are hitched ever more tightly to cybernetic control loops, mind and body subsumed into complex assemblages that render the locus of agency increasingly diffuse and uncertain.[46] Most dramatically, Virilio speaks of an "endo-colonization" in which "the human body is eaten up, invaded, and controlled by technology."[47] Functionally harnessed to its ends, *Homo militaris* is as much the object of the martial gaze as its subject.

Qualifying the Argument

Before proceeding further, a few important qualifications regarding the book's argument are needed. In particular, the central contention of an emerging order of global targeting, in which visibility is synonymous with vulnerability, should not be taken as an endorsement of the overblown and self-serving claims routinely made about the capabilities of military technology. In the last couple of decades, governments, the leaders of their armed forces, and the defense contractors supplying them have keenly embraced and promoted visions of surgical strikes that infallibly target enemies with pinpoint accuracy while sparing the lives of surrounding noncombatants. In addition to fueling delusional fantasies of frictionless exercises of power through military force, the high precision of modern weapon systems is often questionably invoked to assert ethical superiority by the side that uses them.

From the outset, we must forsake any simplistic narrative of a linear progression of increasing accuracy and precision in the actual exercise of war. Indeed, military planners have on many occasions willfully chosen the massing of destructive force over wide areas above any efforts toward target discrimination. As Donald MacKenzie showed in his classic study of nuclear-missile guidance, as long as Cold War nuclear strategists focused on the destruction of large urban areas, accuracy and precision were low technological priorities.[48] More generally, where collateral damage to unintended targets is not a primary concern, the increased destructive range of a munition can compensate for targeting inaccuracy. The pursuit of precision targeting in war is therefore not the preordained outcome of technological evolution and has been conditioned by civilian and military decision makers' political and strategic choices. In particular, the American embrace of precision weaponry in the past few decades has occurred in the context of a decline in mass armies and an increasing public aversion to casualties.[49]

Moreover, the exalted rhetoric of precision targeting should not be mistaken for its reality. Historically, the actual battlefield performance of machines of war has almost always failed to live up to the elevated promises made by their inventors and supporters, sometimes disastrously so. Today is no different. As a result of human error, technical malfunctions, or merely their inherent margins of operational error, the latest weapon systems still can and do miss their intended targets.[50] Collateral damage to noncombatants persists, whether its individual occurrences are construed as the product of unforeseeable mishaps or of a deliberate risk calculus.[51] Intelligence failures and the resulting misidentification of targets remain perennial problems, all the more so when adversaries adopt effective countermeasures of concealment, disguise, and decoying. However sophisticated the available means of surveillance and data acquisition, military intelligence rarely presents itself as entirely free of ambiguity or uncertainty. Independent assessments of the information's credibility have to be made, along with judgments as to the identity or threat of any perceived entity within the battlespace. For all these reasons, a sternly critical eye needs to be cast on any facile self-justificatory claims of an ethical amelioration in the conduct of war provided by precision munitions.[52]

The enduring belief that the acquisition and use of precision weaponry and its accompanying arsenal of perceptual technologies can all but guarantee military superiority and the effortless exertion of political will is no less problematic. The meager outcomes of the various military interventions and other campaigns of targeted killing of the twenty-first century should swiftly dispel any such conceit. The proponents of the latest revival of counterinsurgency doctrine have notably insisted on the limits of mere kinetic force when success is deemed to pass through winning hearts and minds.[53] Yet no matter the reversals and dashed hopes, the quixotic quest for the perfect weapon, so alluring in its promises of invulnerability and unbridled mastery, seems certain to continue.[54]

Although maintaining a critical distance to inflated assertions of contemporary weaponry's ethical and military superiority is essential, certain indisputable trends need to be acknowledged. For instance, there is no denying the long-term increase in the precision of weapon systems over ever-greater ranges of fire. The most common military measure of a weapon system's precision is the Circular Error Probable (CEP), which refers to the radius of a circle around the target within which 50 percent of strikes are expected to fall. Over the past seventy years, the CEP of aerial munitions has seen a marked reduction from 1,000 meters during World War II to 300 meters in the Korean War, 120 meters during the Vietnam War, and presently down to single digits in ideal conditions.[55] While it might have taken on average around nine thousand bombs to hit a given target from the air during World War II, one or two guided munitions are generally sufficient to accomplish a direct strike today. This dramatic alteration in the underlying economy of force evidently has far-reaching ramifications for the conduct of war in toto.

Incontrovertible as the manifest tendency toward increased precision and perceptual reach has become in recent times, it does not thereby follow that it obeys a grand historical telos. In particular, we should eschew the notion that it must logically culminate in a supreme, death-dealing eye spanning the globe. For one, the logistics of military perception is replete with historical contingencies, spurred in particular by the scale and intensity of the past century's violent paroxysms. Moreover, the logistics' practical implementation in concrete battlespaces attests to inherently fragmentary ad hoc processes that never coalesce into a single overarching mechanism. In considering the role of aerial and satellite imagery systems in military combat situations, Chad Harris highlights how the amalgamation of various machines, personnel, institutions, technical standards, and protocols is intrinsically heterogeneous and unstable. The continuous addition of new capabilities and functions, the adjustments being made for the demands of specific operations, and the subsequent improvised work-arounds entail a constant turnover and mutability in these assemblages. These ever-shifting articulations preclude the coherence of "a single, integrated master system" that could "provide perfect transparency for military commanders" through the realization of a comprehensive synoptic overview. Instead, we are always encountering an "architecture contingent on particular historical and political conditions, always in a state of continuous emergence as a series of both institutional and ad hoc structures."[56] There is not, nor is there ever likely to be, a singular unitary gaze surveying the totality of the battlespace. There is no one orb to rule them all but rather a proliferating swarm of unblinking eyes.

A Machinic History

Rather than designating any singular entity or all-embracing totality, the notion of a martial gaze advanced in this book refers to the general disposition and various sociotechnical means accreted toward the rational organization of perception for the ends of military domination and control. In terms of method, the account proposed here is perhaps best characterized as a machinic history, to be distinguished from a mere history of machines.[57] A concern with the designs, material supports, and operational principles of specific technical objects is certainly one of its constitutive elements. No less important, however, is the relating of these devices back to the wider machinic collectives in which they are inserted. Indeed, it is through their entanglement with bodies, knowledges, organizational diagrams, and institutional frameworks that technological artifacts cohere, function, and persist. So when Bernard Stiegler tells us that "the *question of war* is inevitably contained within the *question of technics,*" it does not suggest that only the "technical tool" should interest us. For it is "technicity, as a system, [that] constitutes the artificial and social dispositif of predation and defence from the beginning of humanity."[58] Technology cannot be simply reduced to its material concretions, all the more as it becomes embedded in a scientific apprehension of the world. Following Lorenzo Simpson, technology in its modern guise must refer to "that cumulative body of knowledge, skills, devices and instruments that, using either scientific theory or method, offers us guidance in the selection of courses of action which are adequate for the realization of a given end or for the solution of a given problem and, further, places at our disposal the means to that realization."[59] Not only are technical objects in themselves materializations of congealed knowledge, but their production, deployment, and use also rely on broader collective ensembles.

Such ensembles are referred to in this work as assemblages, inspired by their conceptual elaboration in the philosophy of Gilles Deleuze and Félix Guattari.[60] Assemblages are heterogeneous collectives of entities that cut across the organic, mechanical, and social strata of reality. Within a given assemblage, the machinic denotes the way that it works, as well as how its various elements relate to each other to form a functional, if temporary, whole. Congruent with this conception is the view that the social and technical are inherently hybridized domains, that human groupings are only possible through the material objects that hold them together. In Bruno Latour's words, "We are socio-technical animals and each human interaction is socio-technical."[61] This is not to say that we should treat technical objects as neutral conduits of human agency or as mere conveyer belts for the transmission of wider social forces. Instead, they exert their own form of material agency, shaping and constraining their milieu and the other entities that enter into relation with them.[62] As transient units of becoming, technical objects are stabilized by their degree of internal coherence and the strength of their external relations, but ultimately they evolve in codetermination with the wider assemblages they participate in.[63] Nonetheless, their relative permanence makes them particularly legible snapshots in time, capturing in their physical form the concatenation of intangible forces that gave rise to them.[64]

The present study therefore not only traces the broad arc of socio-technical development but also stops to consider in greater detail some of the specific apparatuses that populate this history. These individual devices and their respective operational influences vary considerably. Some were among the major military technologies of their day, seeing widespread deployment in the field and playing vital roles in contemporaneous conflicts. Other devices are largely forgotten today: speculative designs that were ahead of their time, technological dead ends rendered obsolete or impractical, or simply irredeemably flawed designs that were abandoned. Still others are areas of current research and development that may tomorrow transform warfare or alternatively vanish with barely a trace, perhaps only to resurface in a different guise decades from now. Yet each of these apparatuses counts as a machinic concretion, conceived to fulfill a precise function within wider martial assemblages and possessing its own individual mode of operability derived from the technoscientific repository available in its time. In every case, particular attention will be paid to the specific articulations of organism and machine, drawing out "the situated, heterogeneous couplings of the human and the nonhuman in their visible performativity."[65]

The text is essentially structured around the functional constituents of the martial gaze—sensing, imaging, and mapping—followed by an examination of the obstructive activity of hiding. Given their broad historical and conceptual sweep, these individual accounts cannot possibly be exhaustive. As such, each chapter aspires to offer a representative cut in the historical record, illuminating the general evolution in the prevailing techniques, rationalities, and practices within these domains and highlighting the major accelerations, transitions, and disruptions that have marked their development until the present day.

By way of necessary prelude, this book begins with a genealogy of the intertwined processes of rationalization of vision and mathematization of space. These processes took off in the early modern era and laid the groundwork for the subsequent expansion of the martial gaze. The chapter's main focus is on the technique of linear perspective in the Renaissance's visual arts, as pioneered by Filippo Brunelleschi and theorized by Leon Battista Alberti during the Italian quattrocento. At the intersection of the fields of optical science, pictorial representation, and land surveying, linear perspective's paramount significance is its establishment of a mathematical correspondence between subjective visual perception and objective physical space. These geometrical procedures in turn provide the basis for the subsequent mechanization and automation of vision and imaging so central to the present-day logistics of military perception. While it would take another few centuries for their complete mathematical formalization under a general projective geometry, early technical supports of these procedures are encountered in Albrecht Dürer's ingenious mechanical aids to the painter and other contemporaneous survey instruments.

The following chapter, "Sensing," charts sensorial perception's progressive expansion across the battlespace and its intimate relationship with the targeting of lethal force. An array of devices of varying familiarity feature in this history: rifle scopes, optical range finders, trench periscopes, tachometric bombsights, night-vision goggles, acoustic locators, laser-guided bullets, television bombs, and even pigeon-guided missiles. Across the period covered, we follow the constitution of four successive orders of targeting—aiming, ranging, tracking, and guiding—and their respective regimentations of organic perception. With each order, the distances and accuracy for the projection of lethal force increase, while human assimilation into weapon systems deepens. Moving beyond the original realm of optical magnification, militaries progressively annex the entire electromagnetic spectrum through the deployment of infrared and radar, along with enlisting sonic frequencies via sonar and terrestrial sound ranging. Finally, the advent of the laser promises a singular weapon that can realize the definitive conjunction of perception and annihilation. Tellingly, the laser's first offensive application is to blind the human eye that gave rise to it.

The third chapter, "Imaging," turns its attention to the sociotechnical means by which military organizations record, disseminate, and interpret acts of perception. The photographic camera invented in the mid-nineteenth century is the paradigmatic technology of this component of the martial gaze. Alongside it are found an assortment of camera guns, stereoscopes, reconnaissance aircraft, surveillance satellites, computer vision systems, helmet-mounted displays, and virtual reality control interfaces. Through its vertical elevation above the earth's surface and eventually into planetary orbit, the camera's relentless capture of images becomes an essential component of war from World War I onward. At the close of the last century, two distinct yet interrelated developments gather speed with the introduction of digital imaging bolstered by high-speed algorithmic processing. On the one hand, various computational apparatuses autonomously perform image interpretation and corresponding decision-making tasks. On the other, human operatives are ever more fully immersed within virtual digital environments in which imaging tends to overlay their entire phenomenal field.

Put simply, the rise of machine vision is accompanied by the proliferation of synthetic visions.

Under the heading of "Mapping," chapter 4 considers the crucial function of geospatial representation and geolocalization in war. Along the way, we encounter a plethora of military assemblages that include creeping barrage maps, hyperbolic radio navigation, satellite geopositioning, subterranean sonic mapping, gyroscopic inertial guidance, and terrain contour-matching cruise missiles. Although maps have long assisted the orientation of troops and maritime navigation, their modern scientific form, with its Cartesian graticule of coordinates, acquires a central role in armed conflict. A meticulous spatial striation of the planet is progressively realized, from the gridded artillery maps of World War I to the global coordinate framework of the U.S. Military Grid Reference System. Today, any entity or location can be assigned a precise and unique position on the globe and thus rendered a target. Furthermore, vehicles and weapons can orient themselves and navigate autonomously by reference to digital repositories of geographic information. Indeed, in the closing decades of the twentieth century, the traditional map transmutes into the geographic information system, departing from the form of static cartographical representation toward modes of dynamic visualization of continuously updated geospatial databases. Through the closed circuits of information that bind them ever more tightly, human and computational powers of geographic reasoning irresistibly converge.

The final chapter, "Hiding," turns to the practice of hiding and the various stratagems devised to evade, misdirect, or blind the martial gaze. Systematic concealment and disguise of military units and installations becomes indispensable as the surveillance of the battlespace intensifies. A range of fundamental techniques seeking to blend, disrupt, screen, masquerade, or confuse were pioneered in World War I, yielding khaki uniforms, cubist camouflage patterns, dazzle ships, smoke barriers, counterfeit trees, and imitation corpses. The subsequent acquisition of sensors across the electromagnetic spectrum and diffusion of computational resources generate in turn a new raft of evasive techniques, from infrared-reflecting pigments, stealth aircraft, and algorithmically generated patterns to cloaking metamaterials and active camouflage systems that dynamically adapt to their surroundings. Decoys offer another type of ploy in purposefully drawing the martial gaze to the simulated perceptual signature of a sought-after entity. Deployments include flares, chaff, inflatable tanks, dummy airfields, bogus landmarks, and even a fake Paris. Yet for all these technical countermeasures, a more radical response to the extension of military perception is also presently coming to the fore. Abandoning the bounded physical and normative spaces of the battlefield, various actors are adopting strategies of generalized concealment, dispersal, and expendability that critically erode conventional distinctions between war and peace.

The book concludes with the fateful breakdown of war's distinct spatiotemporality, as induced by the combined action of global targeting and total dissimulation. Submitted to unremitting processes of domestic securitization, societies whose armed forces and national defense establishments have propelled the martial gaze's ascendancy now find themselves increasingly caught within its scrutiny, their immunity from its correlated targeting rendered ever more precarious. Yet most perilous of all is arguably the fate of a vacillating human subject whose putative autonomy dissipates as it is increasingly embedded within a logistical order of military perception.

CHAPTER 1

Perspective

> And thus the eye is the ruler and the compass of distant regions and of longitudes and abstract lines. Everything is comprehended under the geometric doctrine, and with the aid of the arithmetic art we see that there is a rule for measuring . . . with the eye.
>
> —GIOVANNI CAVALCANTI, *Istorie Fiorentine* (1440)

Heralded as "the most important event of the Renaissance," the invention of linear perspective occupies a crucial role in the intertwined process of rationalization of vision and mathematization of space that constitutes one of the defining features of historical modernity.[1] Much more than a mere internal development within the history of art, linear perspective emerges at the confluence of the fields of geometrical optics, pictorial representation, and land surveying. As such, perspective is at the root of the unfolding of the martial gaze's three interdependent functions of sensing, imaging, and mapping. At its heart is a geometric conception of vision that accounts for the natural ocular estimation of spatial distances and begets an array of mathematical techniques and associated instruments for precise measurement by the human eye. A rigorous procedure for the rule-governed (i.e., automatable) construction of images that convey an optically convincing depiction of physical space follows from it. In common with the field of cartography taking shape at the same time, perspective uses a system of geometric projection of a three-dimensional space on a two-dimensional surface in a manner that preserves the relative proportions of that space and the objects contained within it.

In ordering visual space according to an abstract system of linear coordinates, perspective binds subjective perception and objective spatial extension through mathematical convertibility. For William Ivins, perspective thus offers a "practical means for securing a rigorous two-way, or reciprocal, metrical relationship between the shapes of objects as definitely located in space and their pictorial representations."[2] Bruno Latour further points to the paramount importance of the "optical consistency" found in Renaissance techniques of visualization for the development of modern scientific culture. Common to linear perspective, scientific mapping, and technical drawings, optical consistency permits the description of physical spaces and objects in a systematic geometric fashion that facilitates the circulation of these representations and allows for "the possibility of going from one type of visual trace to another."[3] According to Latour, these particular representations possess the character of "immutable mobiles" able to disseminate widely and undergo rule-bound transformations and additions without losing or distorting their valuable informational content. These new itinerant forms of inscription not only serve as repositories of accumulated knowledge but as material anchors for the cognitive operations necessary to the emergence and sustenance of a modern scientific culture.[4] While these properties do not flow exclusively from it, linear perspective is nonetheless paradigmatic by virtue of "its logical recognition of internal invariances through all the transformations produced by changes in spatial location."[5] Furthermore, although optically consistent visualizations have decisively contributed to the success and proliferation of scientific practices, they can also be considered more generally as instruments of power for shaping the natural and social world. Indeed, for our purposes, we can identify their especially important role in the extension and coordination of military operations across ever-greater geographic expanses, including the remote projection of weaponry. We can hence approach linear perspective as one of the foundational sites of the technoscientific tethering of perception to the imperatives of targeting.

From the outset, perspective participated in an objectivization of perception, no longer solely residing in phenomenal embodiment but instead now the outcome of a systematic application of "a rational and repeatable procedure."[6] Moreover, this algorithmic apprehension of perception made possible its comprehensive mechanization and repurposing within broader sociotechnical assemblages. Neither perspective nor its associated machines faithfully replicate human vision, of course. To the roving, binocular, and psychologically conditioned sight of humans is substituted a fixed, monocular, and strictly geometric gaze. Yet it is the disembodied and rationalized conception of perception that increasingly prevails and imposes itself on the human sensorium. As Ken Hillis puts it, "The body-as-object (and by extension humanity) becomes subject to, even inscribed by, the laws of technique that are intended to render nature more subject to human control."[7]

This chapter begins by charting the historical emergence and significance of linear perspective, outlining its fundamental principles and technical precepts as established by its initiators. We then turn to the variety of technical devices invented to assist in the production of perspectival images, foreshadowing the wholesale automation of perception and its representations to come in subsequent chapters. Finally, we consider linear perspective's progressive assimilation into a generalized projective geometry, which encompasses previously distinct graphical projections such as the orthographic soldier's perspective associated with engineering and architectural plans. This mathematical formalization of perspective geometry effectively completes the rationalization of vision initiated in the Renaissance and remains to this day a major constituent of the logistics of military perception.

The Rationalization of Vision

In precise terms, linear perspective can be defined as the procedure for making "a picture on a flat surface in such a manner that the various objects represented in it appear to have the same sizes, shapes, and positions, relatively to each other, that the actual objects as located in actual space would have if seen by the beholder from a single determined point of view."[8] Typically these images possess a single vanishing point toward which all the parallel lines perpendicular to the picture plane converge, presenting a convincing illusion of depth perception to the viewer.[9] Universally acknowledged as a landmark event in the history of art, the advent of linear perspective during the Italian Renaissance has bequeathed us some of the most influential and celebrated paintings in the Western canon. Yet beyond the artistic import of these creations, their geometric constructions participated in a dual process of rationalization of vision and mathematization of space, the social and cultural ramifications of which are wide ranging and profound.

From the first demonstrations of Filippo Brunelleschi in the early fifteenth century to the subsequent systematizations of Leon Battista

Alberti and Albrecht Dürer, perspectivists developed a set of methods and technical aids for the production of images that simultaneously bear optical verisimilitude to embodied vision and accurately record the relative positions and dimensions of objects within a homogeneous space. For James Burke, the impact of the underlying perspective geometry was nothing short of revolutionary: "The new technique permitted the world to be measured through proportional comparison. With the aid of the new geometry the relative sizes of different objects could be assessed at a distance for the first time. Distant objects could be reproduced with fidelity, or created to exact specifications in any position in space and then manipulated mathematically."[10] We cannot understate this procedure's central role in the burgeoning scientific worldview. William Ivins notably underlines that "science and technology have advanced in more than direct ratio to the ability of men to contrive methods by which phenomena [. . .] have been brought within the range of visual recognition and measurements and then become subject to that logical symbolization without which rational thought and analysis are impossible."[11]

The history of linear perspective and its origins has been the subject of a vast and rich literature, which is neither possible nor useful to fully render here. The appearance of perspective has been variously explained, whether by reference to the advent of a new "worldview" or to practical developments in architecture, workshop practice, surveying, cartography, and astronomy.[12] Historians have argued persuasive cases for each of these explanations, but no clear consensus exists for the most significant or decisive one. For the purposes of our study, an independent account of the origins of perspective or a definitive adjudication between the various existing claims is not necessary. It suffices merely to draw out the mutual dependencies and common underlying logic weaving these various fields of activity together. Indeed, the very entanglement of the geometric methods used across all these domains strongly suggests that there is no single origin to the development of perspective. From this entanglement, we may further conclude that the functional relations binding the modern techniques and machines of sensing, imaging, and mapping were already implicit at their beginnings.

Whether it originated or simply cohered alongside the creation of linear perspective, a powerful new worldview certainly came into focus, imbuing a universal and homogeneous spatial geometry with profound intellectual and cultural significance. If classical texts of the ancient world such as Ptolemy's *Geographia* (freshly rediscovered by Western

Christendom in the early fifteenth century) and Euclid's *Elements of Geometry* and *Optics* no doubt played an important role in its development, linear perspective nonetheless remains an original invention of the Italian Renaissance. Classical art's geometric representations approximated linear perspective but never gave rise to a full and systematic perspectival depiction of space, despite the ancients having all the required knowledge of geometry. In his landmark essay on *Perspective as Symbolic Form,* Erwin Panofsky explains the absence of any conception of a homogeneous continuous space in the ancient world: "They did not do it because that feeling for space which was seeking expression in the plastic arts simply did not demand a systematic space."[13] It was only over the course of the late Middle Ages and the Renaissance, Panofsky argues, that modern notions of a homogeneous and isotropic "systematic space" emerged and found artistic expression in linear perspective. The move toward perspectival representation in art in turn supported the emergence of the unified physicalist conception of space that underpinned the new empiricism of the scientific revolution and its crowning achievement in Newton's universal laws of motion. In the words of Richard Tarnas, "It was not peripheral to the emergence of the modern outlook that Renaissance art depicted a world of rationally related solids in a unified space seen from a single objective viewpoint."[14] Most important for our purposes, perspective established a mathematical correspondence between objective physical space and subjective visual space, assuming the latter to be "ordered a priori by an abstract, uniform system of linear coordinates" that faithfully reflects the former.[15]

Art historians widely credit Italian architect Filippo Brunelleschi with the first recorded demonstration of linear perspective. Around 1425, Brunelleschi presented a perspectival painting of the Baptistery of St. John in Florence for viewing through an arrestingly intricate setup. The viewer would stand at the exact location where the work had been painted and peer with one eye through a small hole at the back of the painting in the direction of the real baptistery. A mirror would then be raised in front of the viewer, thus reflecting back the painted image. The created illusion was so compelling that reality and its simulacrum were said to be virtually indistinguishable.[16] Although Brunelleschi's exact method for creating this perspectival image is still debated, his demonstration suggests not merely a certain flair for showmanship but also an astute understanding of optical laws from the precise positioning and size of mirror required for the illusion.[17]

Martin Kemp concludes from Brunelleschi's architectural background that his perspective technique likely derived from surveying practices.[18] In the late Middle Ages, practical geometric techniques of surveying, through such instruments as the cross-staff, quadrant, or astrolabe, were increasingly supplemented by the rediscovery of Greek trigonometric theory via the translation and diffusion of Islamic scientific texts. Kim Veltman submits that surveying and perspective are fundamentally entangled because their aims are "essentially the same": "both seek to record objective elements of the measured world."[19] Perspective and surveying are thus equally exercises in measurement by eye such that "a picture in the modern perspectival style is essentially the outcome of a measuring procedure."[20] Kemp has likewise underlined the strength of the "intersection of geometrical perspective and its related machines with the mathematics of mensuration."[21]

The first attempt at formalizing the rules of linear perspectival representation appears in the 1435 treatise *De Pictura* by Brunelleschi's fellow Florentine, Leon Battista Alberti. In common with many of his contemporaries, Alberti granted the eye an exalted status among the senses: "It is such as to be the first, chief, king, like a god of human parts."[22] He explicitly grounded his theory of perspective in optics, and Joel Snyder sees in his treatise "the first effort by a painter to establish the certainty of his method of picture construction by deriving it from a scientific account of vision."[23] Although primarily writing for artists, Alberti insisted that "one who is ignorant of geometry will not understand these or any other rules of painting."[24] In effect, the geometric principles used are the same as those found in surveying. Alberti explains that the eye measures distances (or "quantity") between two points in space with the lines of sight ("visual rays") to each point "as with a pair of compasses." In this way, "vision makes a triangle" in which "the base of [this triangle] is the quantity seen and the sides are those rays which are extended from the quantity to the eye."[25] The more acute the angle "within the eye" (i.e., the angle formed by the two lines of sight with each other), the smaller the distance between the two points being observed will appear, a principle familiar to surveyors that captures the universally recognized inverse relationship between distance to an object and visible size. Together, the combined visual rays constitute a "visual pyramid" (Figure 2). Leonardo da Vinci could thus later present perspective as "the demonstrative proof that is confirmed by

FIGURE 2. *Alberti's visual pyramids. Abraham Bosse,* Manière universelle de M. Desargues pour pratiquer la perspective *(1648). Bibliothèque nationale de France.*

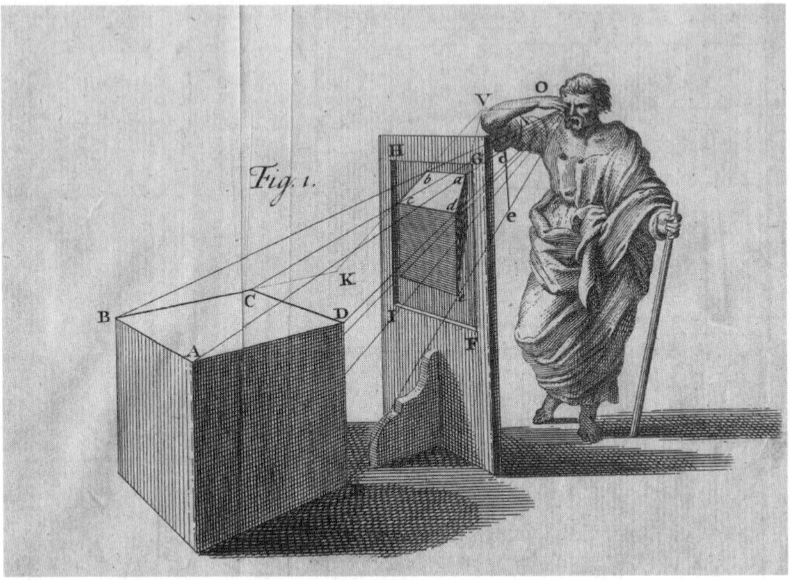

FIGURE 3. *Perspectival image as planar section of the visual pyramid. Brook Taylor,* New Principles of Linear Perspective *(1811).*

experience that all things communicate their likenesses to the eye by pyramidal lines; and that the pyramids of bodies of equal magnitudes form a larger or smaller angle according to the kind of distance that there is between one and the other."[26]

Within this geometric conception of vision, the artist's picture corresponds to a planar section of the visual pyramid at a given distance (Figure 3). For Alberti, a painting is therefore analogous to "an open window through which I see what I want to paint."[27] Realized in conformity with the established principles of linear perspective, the depicted space appears as an immediate extension of the viewer's own physical space. Consequently, there can be only one optically correct position from which to view any such representation, at the very apex of the visual pyramid being intersected by the picture plane. In Alberti's words, "A painted thing can never appear truthful where there is not a definite distance for seeing it."[28] This unique viewpoint is mirrored in the picture by a center point at which the shortest and only perpendicular line of sight—the central ray of the visual pyramid that Alberti dubs "the prince of rays"—arrives on the planar section. In the paradigmatic case of central one-point perspective, the center point coincides with the image's single vanishing point, toward which converge all the lines that are parallel with the viewer's line of sight (or, put in different but equivalent terms, all the lines perpendicular to the picture plane). Brunelleschi's elaborate demonstration had indeed already established "the projective coincidence of the point of view and the vanishing point" through the contrivance of the mirror, the single eye of the viewer being required to peer out from the painting's back panel at the very spot of the vanishing point.[29]

A perspectival representation thus simultaneously fixes the relative positions and proportions of objects in the depicted space and the location from which that scene is being viewed. The great stillness of Renaissance painting is also that of its viewer. The perspectival image is a two-dimensional plane that opens up simultaneously into the imagined three-dimensional space of the depicted scene and into the real physical space of its singular viewer. For David L. Martin, the central visual axis of perspective connecting the vanishing point to the viewpoint accordingly presents the viewer with "an image of the self as a visible, objectified, and measurable unit distributed in space."[30] It is worth pausing here to consider the profound implications of the human viewer's objectivization. Indeed, according to a prominent line of interpretation, linear perspective arranged the visible world "for the spectator as the universe was once thought to be arranged for God," enthroning the human subject as its new sovereign.[31] James Burke asserts, "At a stroke, the special relationship between God and every separate object was removed, to be replaced by direct human control over objects existing in the same, measurable space."[32] It is indisputable that the nascent humanism of the age was accompanied by an increased capacity for control over a desacralized world. However, the notion that this control issued straightforwardly from a sovereign human subject is problematized when we consider that this very subject was being simultaneously rendered as a new object of rational knowledge located in the same homogeneous mathematized space, its embodied perception now subordinated to abstract laws of vision.[33] From this point on, an incremental but inexorable disembedding of perception from the site of the living organism was propelled forward through the instantiation—initially partial and then increasingly comprehensive—of its mechanisms in technical apparatuses. This deracination should not, however, be understood as

FIGURE 4. *Perspectival subject. Hans Vredeman de Vries,* Perspective *(1604). Digital image courtesy of the Getty's Open Content Program.*

merely the material application of an already established mental abstraction. Rather, the technical mediation and appropriation of perception was intertwined with its theoretical conceptualization from the outset.

Perspective Machines

Friedrich Kittler observes that in the "artisanal, hand-crafted phase of optical media" characterized by Renaissance painting, its productions would already have been inconceivable "without calculations, and thus they required a technical-scientific foundation." From this foundation, the "technical apparatus" would develop until it could "detach itself from the eyes and hands of so-called artists and form those absolutely autonomous spheres" that are contemporary "optical media technologies."[34] As William J. Mitchell puts it, Alberti's writings ultimately offered nothing less than "a rigorous, step-by-step algorithm for constructing, with drafting instruments, a consistent single-point perspective view of a scene."[35] It is precisely this "algorithmic character" that has primed perspectival representation for subsequent mechanization and eventually automation.[36]

While Alberti's geometric instructions to the artist were precise and systematic, their application still required a mathematical competence that exceeded that of most contemporary practitioners. Over the next few centuries, an array of techniques and devices were conceived to assist in the production of perspectivally accurate images. Alberti himself had proposed a grid or *velo* (veil) that the artist would place between the eye and the represented object, such that it would constitute a plane intersecting the visual pyramid. By fixing the position of the eye, artists could reproduce the details of the model, as seen through the *velo,* on a similarly gridded surface. Albrecht Dürer's influential treatise on perspective, *Underweysung der Messung* (Instruction of measurement) presents a similar setup in which a gridded mesh is placed between the artist's eye and the scene to be represented. (See Figure 5 for a later depiction of the *velo.*)[37] However, Dürer suggests another technique that is of particular interest: a simple instrument that uses a piece of string to mimic the visual rays from the model to the viewer, the respective positions of which can be transferred onto a screen for every point on the model (Figure 6). Without the requirement of any geometric knowledge, the apparatus permits the methodical projection of individual points in three-dimensional space onto a two-dimensional picture plane.

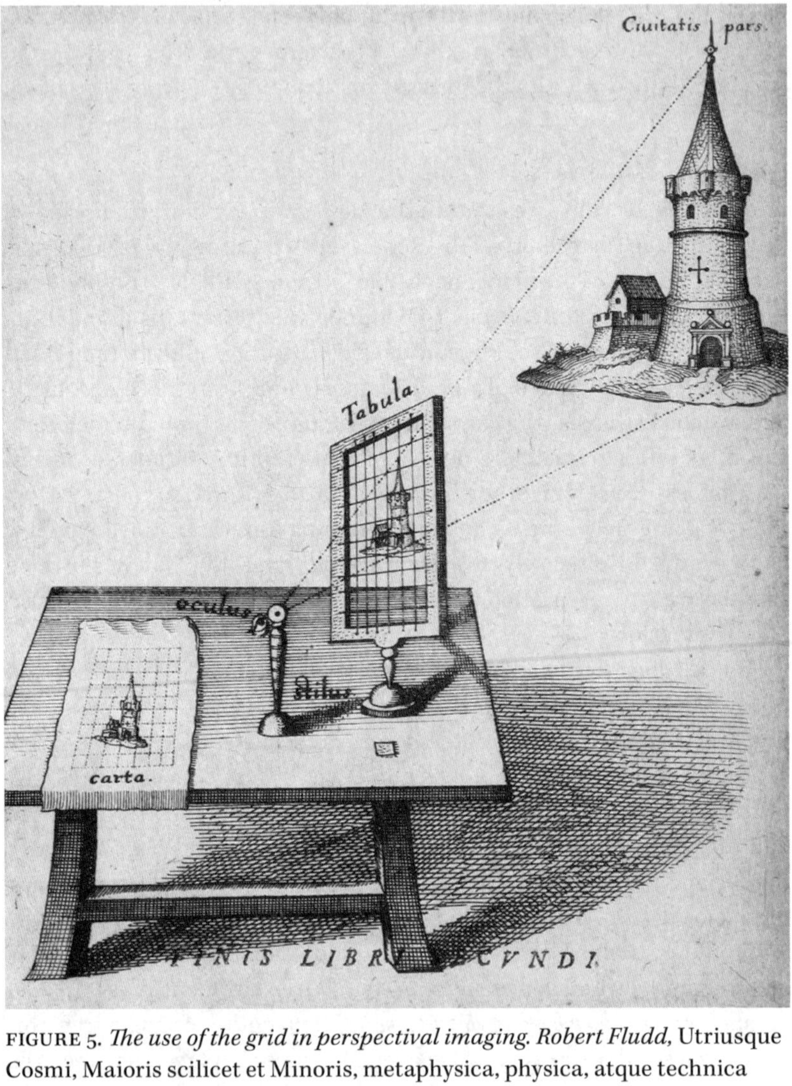

FIGURE 5. *The use of the grid in perspectival imaging. Robert Fludd,* Utriusque Cosmi, Maioris scilicet et Minoris, metaphysica, physica, atque technica Historia *(1617). Digital image courtesy of the Getty's Open Content Program.*

Dürer's device is noteworthy for making explicit the act of measurement involved in the perspectival system of pictorial representation.[38] Perhaps most significantly, it "eliminates the human eye altogether,"[39] thus marking an early instantiation of the efforts aimed at "the mechanization of the perspective painting and of the arrested eye" that would eventually culminate in the invention of photography.[40]

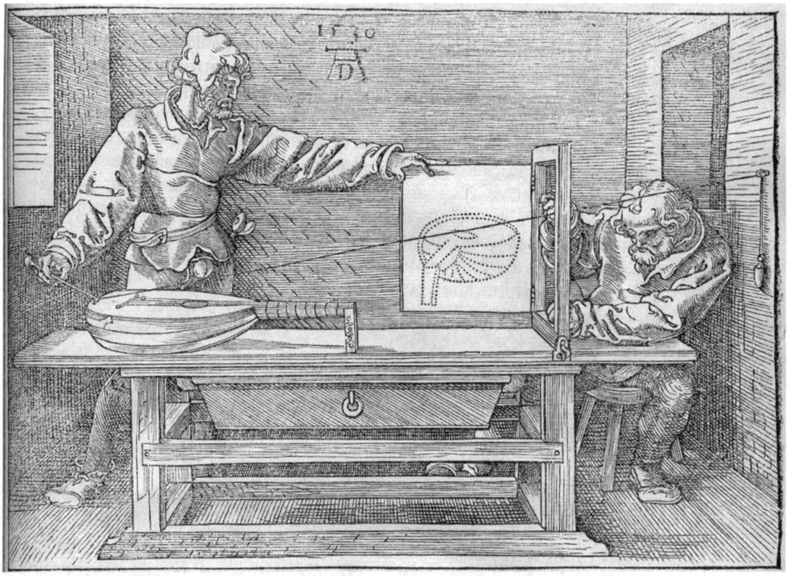

FIGURE 6. *Dürer's string method. Albrecht Dürer,* Underweysung der Messung mit dem Zirkel und Richtscheyt *(1525).*

Such instruments' further development and diffusion were crucial in making perspective accessible to practitioners unschooled in the finer points of geometry and in widening its range of applications, notably in the military field. Thus the introduction to the 1583 edition of Vignola's treatise on "practical perspective" submits that

> the practice of this most delightful art, which besides the many useful things it brings to the military arts, is also particularly important in the construction and defence of forts, since one can with the instruments of these arts draw in outline whatever site one wants without trouble and one will have not only the plane but also the elevation in its every detail and the measure of its parts will be proportionate to the distance which there is between our eye and the things we have measured in our drawing.[41]

Over the next two centuries, numerous works were produced to educate military staff on the practical applications of perspective and geometric drawing, from the design of fortifications and the production of maps to the deployment of troops on the battlefield. As Veltman notes, these

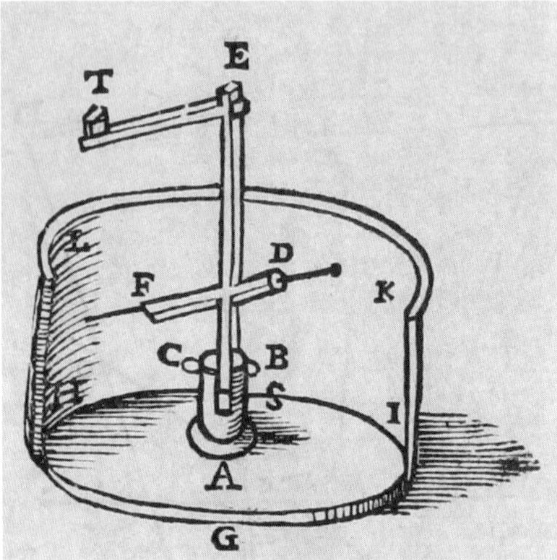

FIGURE 7. *Baldassarre Lanci's* distanziometro, *depicted in Giacomo Barozzi da Vignola,* Le Due Regole della Prospettiva Practica *(1583). Digital image courtesy of the Getty's Open Content Program.*

manuals typically bypassed the underlying geometric theory because a range of contemporary instruments could produce perspectival representations without the requirement of specialist knowledge by the user.[42] Such instruments included variants of grid or string methods and adaptations of the Jacob's staff, quadrant, or astrolabe. Among them was the ingenious *distanziometro* (distance meter) invented by Baldassarre Lanci in 1557. Primarily intended for capturing the layout of fortifications, the device permitted the accurate production of perspectival drawings through the visual surveying of spatial points (Figure 7). For every position sighted through the device's rotating eyepiece, an affixed stylus would inscribe a point on a curved sheet of paper. Once sufficient points had been registered, these could be joined up to form a faithful perspectival representation of the scene surveyed.

While not devoid of aesthetic appeal, such perspectival images were eminently practical. They provided accurate information about the relative proportions and positions of objects in space, which was plainly valuable to the planning and conduct of military operations. Pietro Accolti's 1625 work *Lo Inganno de gl'occhi* (The deception of the eyes) details how, from the perspectival drawing of a fortress, "it is possible to reconstruct the layout of the building by applying the perspective procedure inversely" by means of Dürer's string technique (Figure 8).[43] Reconnaissance missions could, from a safe distance and with the assistance of appropriate instruments, produce a perspectival sketch of

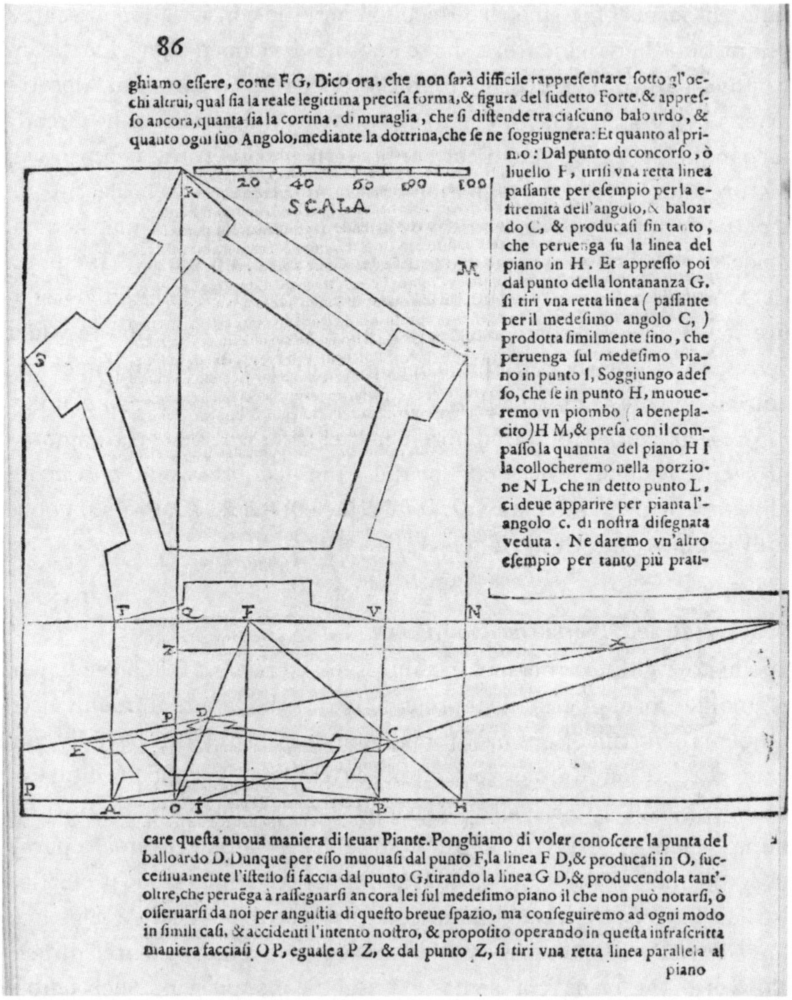

86

ghiamo eſſere, come F G, Dico ora, che non ſarà difficile rappreſentare ſotto gl'occhi altrui, qual ſia la reale legittima preciſa forma, & figura del ſudetto Forte, & appreſſo ancora, quanta ſia la cortina, di muraglia, che ſi diſtende tra ciaſcuno baloardo, & quanto ogni ſuo Angolo, mediante la dottrina, che ſe ne ſoggiugnerà: Et quanto al primo: Dal punto di concorſo, ò buello F, tiriſi vna retta linea paſſante per eſempio per la eſtremità dell'angolo, ò baloardo C, & producaſi fin tanto, che peruenga ſu la linea del piano in H. Et appreſſo poi dal punto della lontananza G. ſi tiri vna retta linea (paſſante per il medeſimo angolo C,) prodotta ſimilmente ſino, che peruenga ſul medeſimo piano in punto I, Soggiungo adeſſo, che ſe in punto H, muoueremo vn piombo (a beneplacito)H M, & preſa con il compaſſo la quantità del piano H I la collocheremo nella porzione N L, che in detto punto L, ci deue apparire per pianta l'angolo C. di noſtra diſegnata veduta. Ne daremo vn'altro eſempio per tanto più praticare queſta nuoua maniera di leuar Piante. Ponghiamo di voler conoſcere la punta del balloardo D. Dunque per eſſo muouaſi dal punto F, la linea F D, & producaſi in O, ſucceſſiuamente l'iſteſſo ſi faccia dal punto G, tirando la linea G D, & producendola tant'oltre, che peruēga à raſſegnarſi ancora lei ſul medeſimo piano il che non può notarſi, ò oſſeruarſi da noi per anguſtia di queſto breue ſpazio, ma conſeguiremo ad ogni modo in ſimili caſi, & accidenti l'intento noſtro, & propoſito operando in queſta infraſcritta maniera facciaſi O P, eguale a P Z, & dal punto Z, ſi tiri vna retta linea parallela al piano

FIGURE 8. *Method for translating the perspective drawing of a fortress into a plan. Pietro Accolti,* Lo Inganno de gl'occhi *(1625). Digital image courtesy of the Getty's Open Content Program.*

a given site from which an accurate survey map could then be drawn up. While the use of perspective in the field was largely restricted to the surveying of fortifications, this application would be of no small benefit to military commanders, given siege warfare's central strategic role in Europe until the late eighteenth century.

The above techniques underscore that a practical understanding of different forms of geometric projection and the conservation of spatial proportions in moving from one to another developed alongside the invention of linear perspective. Joan Gadol notably argues that Alberti's extensive work on surveying and cartography indicates that he already recognized a "strict, technical connection" between the perspectival picture and the map through their common reliance on a projective geometry designed to preserve the relative distances and dimensions of objects in their translation onto a two-dimensional plane.[44] The practical methods devised in the early modern period undoubtedly prefigure the systematic correlation of situated acts of perception to wider geospatial frameworks that has become indispensable to the machinic automation of the martial gaze's combined functions of sensing, imaging, and mapping. Yet the comprehensive formalization of the common projective geometry that underlay these initial efforts would only be realized through an incremental intellectual enterprise that would not be fully completed until the nineteenth century.

A Generalized Projective Geometry

The history of linear perspective intersects here with the development of another form of graphical projection, one that was commonly used for plans of fortifications or war machines because of its especially convenient representation of spatial measurements. So-called military or soldier's perspective is a form of geometric projection with no vanishing point, where all the parallel lines of the depicted object remain parallel in the image. The value of such a parallel or orthographic projection is that it avoids the visual distortion of classical perspective known as foreshortening. Instead, the projection maintains the measured dimensions of all the forms represented at a single constant scale. Such a property is of particular practical value to architects and engineers, notably for the representation of fortifications or war machines (Figure 9). The use of this graphic form, although frequently lacking in geometrical rigor, preceded the Renaissance discovery of linear perspective and continued to develop primarily within the distinct literature of technical drawing.[45] Yet the linear and military perspectives would eventually be brought together when it was grasped they could both be treated as merely variants of a wider projective geometry. Military perspective can thus be defined as a zero-point perspective that produces an image in

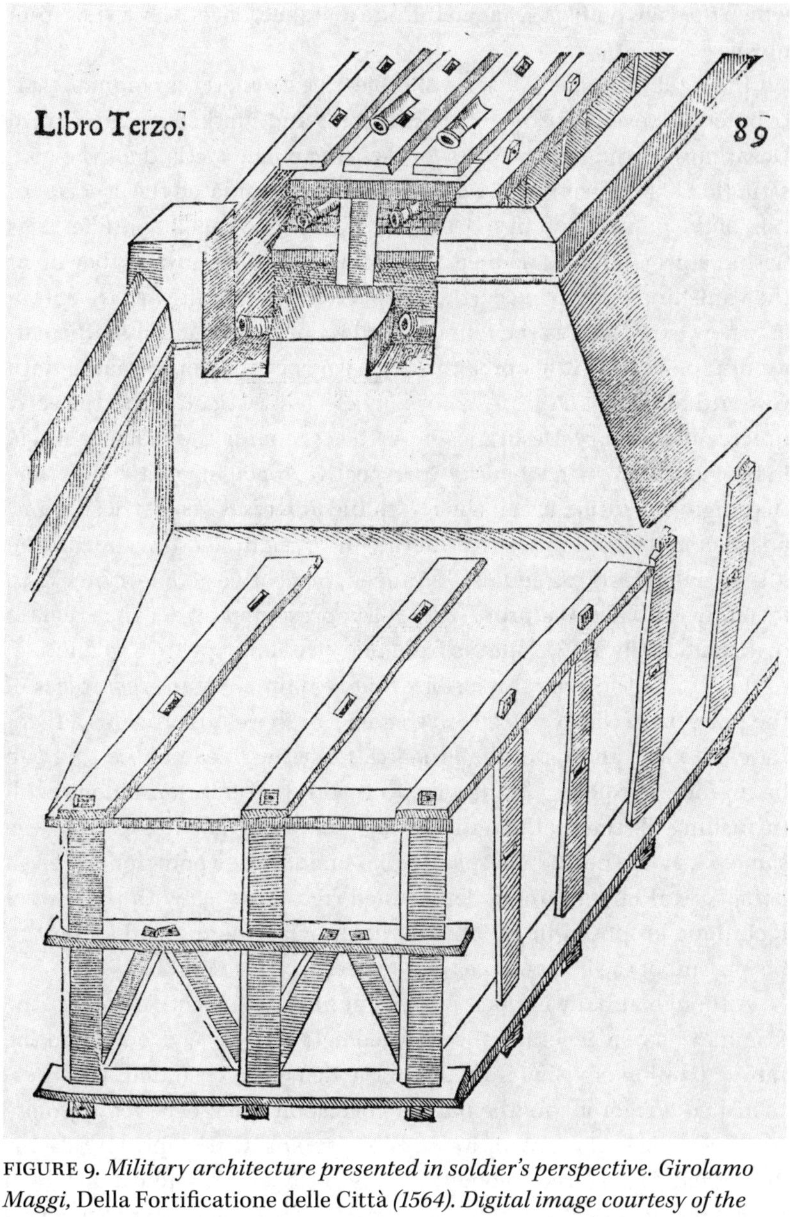

FIGURE 9. *Military architecture presented in soldier's perspective. Girolamo Maggi,* Della Fortificatione delle Città *(1564). Digital image courtesy of the Getty's Open Content Program.*

which the viewpoint is set at an infinite distance, effectively a view from nowhere.

The establishment of a general projective geometry is commonly attributed to seventeenth-century architect and mathematician Girard Desargues. Through his efforts for devising a new method for the construction of perspectival images, Desargues formulated the crucial notion of the point at infinity. Desargues postulated that all parallel lines within a projective plane meet at a point at infinity, just as they do at the vanishing point of a central perspective image, but with this property now expressed as the founding axiom of a generalized mathematics of projection.[46] If his breakthrough was not fully appreciated until Gaspard Monge's and Jean-Victor Poncelet's investigations in the early nineteenth century, Desargues nevertheless marks the moment when, in the words of Erwin Panofsky, "perspective, replacing for the first time the single Euclidean 'visual cone' with the universal 'geometrical beam,' abstracted itself completely from the line of sight and thus opened up all spatial directions equally."[47] Indeed, projective geometry seeks to formally establish the properties of a represented object that remain invariant as the viewer moves from one visual perspective to another, to determine the spatial relations that remain constant regardless of the position of the eye. Projective geometry therefore provides a theoretical understanding of the properties of a Euclidean space that can be extracted from any of its perspectival representations, along with furnishing all the mathematical tools necessary for doing so. By the same token, the position in space of the singular viewpoint for any given perspectival image can be determined rigorously, as with the formal technique proposed in 1759 by mathematician Johann H. Lambert in his pursuit of the inverse rules of perspective.[48]

Although already implicit in the first linear perspective images, the formulation of a general projective geometry effectively completed the rationalization of vision initiated in the Renaissance. It definitively established a rigorous mathematical correspondence between absolute physical space and any of its possible perspectival representations. The rules of projective geometry would provide a standard by which the spatial fidelity of visual perception could be gauged, simultaneously displacing and devaluing innate phenomenal experience. In this spirit, seventeenth-century artist Abraham Bosse, a student under Desargues and a lifelong promoter of his ideas, would affirm that painting's true task was not to represent things "as the eye sees them or believes it sees them, but such as the laws of perspective impose them on our reason."[49]

In the present day, the branch of projective geometry that has incorporated linear perspective underlies the techniques used in fields as varied as photogrammetry, cartography, computer vision, and 3-D graphics. Perspectival geometry thus supports the martial gaze through its deployment within an array of sociotechnical assemblages that seek to visually record, measure, and represent the relative positions and proportions of objects in space. This is not to say that perspective still occupies the sole or even primary role in the contemporary organization of military perception. In many instances, it has given way to other perceptual techniques that do not rest on classical optics and that extend across the electromagnetic spectrum of light, such as radar, infrared, or laser. However, linear perspective has been critical in founding what Lev Manovich refers to as "visual nominalism," understood as "the use of vision to capture the identity of individual objects and spaces by recording distances and shapes" and to "create detailed maps of three-dimensional reality."[50] If a perceptual nominalism now extends far beyond the narrow confines of visible light and accordingly relies on different sets of scientific principles, it remains nonetheless fundamentally tributary to that originary geometric correlation of vision and space. Most fateful of all, the discovery of linear perspective set in motion an autonomization of perception and spatial reasoning, the technical realization of which is still unfolding today before our humbled eyes.

CHAPTER 2

Sensing

> Once we have seen something, we have already started to destroy it.
>
> —PAUL VIRILIO

For millennia, the human body's innate sensory faculties have been the principal perceptual guide to war. Although hearing and smell played their part, it fell to the light-sensitive eye to supply the preeminent martial sense of vision. Weapons used in close-quarters combat demand hand–eye coordination, whether these are simply the body's forelimbs or the myriad clubbing, cutting, or piercing appendages devised to enhance their injurious potential. A weapon's offensive action requires its orientation, projection, and continuous guidance by reference to the relative position and motion of the intended target. At close ranges of engagement, the information necessary for this purpose is obtained primarily, if not exclusively, via the corporeal visual system and acted on through a combination of instinct and prior training.[1] Where such intimate weapons still play a role in combat—assuredly a peripheral one in the present era—this embodied, spontaneous function of vision persists essentially unchanged.

In contrast, a wider survey of the eye at war today reveals it to be systematically solicited, augmented, and assimilated within an extensive array of sociotechnical assemblages. Alongside this mobilization and corralling of the human organism, the visual faculty has itself become increasingly disembedded from its original biological substrate through the incremental rationalization and mechanization of its functions. A primary motive force of this trajectory lies in the development of projectile weaponry capable of striking entities at a distance, a trend particularly accelerated by the advent of gunpowder. The ambition of this chapter is precisely to trace the history of the extensions and reorganizations of sense perception as it has become ever more firmly tethered to the operations of remote targeting.

Our account opens at the turn of the seventeenth century, with the fateful unveiling of the optical telescope and its striking magnification of distant objects from the refraction of light through curved glass lenses. We follow the growing importance of optical apparatuses in war thereafter, markedly spurred by the dramatic increases in the range and accuracy of weaponry inaugurated in the late nineteenth century. The chapter then turns to the specific articulations of vision realized within the successive orders of targeting constituted by aiming, ranging, tracking, and guiding. A systematic outline of their operational principles and particular material instantiations will detail how each new order recasts perception in an ever-deepening imbrication with weaponry. While the initial martial investment of the perceptual field takes place primarily within the narrow band of light visible to the human eye, sensorial reach presently extends well beyond such organic strictures. The infrared range of electromagnetic frequencies occupies a vital role in modern military operations, overcoming nocturnal obscurity and discerning radiant heat sources. Similarly, radar technologies, which harness invisible radio waves to record the presence and motion of objects in space over vast distances, have profoundly altered the conduct of war. Beyond the electromagnetic spectrum, the properties of acoustic phenomena have similarly been marshaled by technologies of sound ranging or underwater sonar. Finally, laser's stimulated emission of photons not only provides further means of sensing the environment and the entities within it but also holds the promise of weaponizing light itself, thereby fulfilling the definitive alignment of perception and annihilation.

Through a Glass, Deadly

Attribution of the invention of the telescope has long been the source of heated debate among historians of technology, but such disputes over its precise origins need not concern us here.[2] In effect, the telescope was the cumulative outcome of experiments with glasswork and mirrors, and enquiries into the laws of optics that go at least as far as back as antiquity.[3] Certainly, by the second half of the thirteenth century, vision-correcting eyeglasses were being put into use as an understanding of the optical magnification of distant objects through curved lenses grew. Medieval philosopher Roger Bacon outlined the optical principles for a practical telescopic device and had already identified potential military usages: "We can so shape transparent bodies, and arrange them in such a way with respect to our sight and objects of vision, that the rays will be refracted and bent in any direction we desire, and under any angle we wish we may see the object near or at a distance. . . . Thus a small army might appear very large, and situated at a distance might appear close at hand."[4] Bacon further asserted the dubious claim that Julius Caesar possessed similar "great glasses" that enabled him to view British fortifications from Gaul, a variation on a number of apocryphal tales, prevalent in the Middle Ages, about telescopic marvels of the ancient world.[5] Such unsubstantiated accounts are less interesting for any veracity they might have than in underscoring the enduring appeal of telescopic vision and the importance accorded to its military uses over the more familiar astronomical applications chronicled in popular histories.

Speculative claims according to which actual magnifying devices were in the possession of Bacon or other early writers who expressed familiarity with their underlying optical principles remain uncorroborated. Indeed, we do not find incontrovertible evidence of a practical telescope until the dawn of the seventeenth century. In September 1608, a Dutch spectacle maker named Hans Lippershey demonstrated a telescopic device at the court of Count Maurice of Nassau, stadtholder of the provinces of Holland and Zeeland and among the most distinguished military leaders of the age.[6] The new apparatus made an instant impression on the enthralled audience, prompting Maurice to quip that "with those glasses they would see the tricks of the enemy." The visiting commander of the Spanish forces in the Low Countries, Ambrogio Spinola, was equally struck, portentously exclaiming, "From now on I can no longer be safe, for you will see me from afar." His hosts seemed to concur, but they charitably promised that they would forbid their men to shoot at him.[7]

As news of its remarkable properties spread across Europe, Lippershey's telescope soon made its way into the hands of the authorities in France, Rome, and the Iberian Peninsula. The device's military applications were immediately evident to its early users. Reporting back to the Vatican in 1609, the pope's envoy in Brussels, Guido Bentivoglio, wrote that Maurice had "procured this instrument in order in time of war to reconnoitre from a distance, and observe places he might want to besiege, or sites of encampments, or enemy forces on the march, or similar situations that might be turned to his advantage."[8] Initially skeptical of the first reports of the telescope, Galileo Galilei would soon seize the opportunity to fashion his own device, offering it to the doge of Venice in a letter addressed to him in 1609. Emphasizing its military capabilities rather than the astronomical utility that would later establish his fame, Galileo wrote of

> a new contrivance of glasses, drawn from the most recondite speculations of perspective, which render visible objects so close to the eye and represent them so distinctly that those that are distant, for example, nine miles appear as though they were only one mile distant. This is a thing of inestimable benefit for all transactions and undertakings, maritime or terrestrial, allowing us at sea to discover at a much greater distance than usual the hulls and sails of the enemy, so that for two hours or more we can detect him before he detects us and, distinguishing the number and kind of vessels, judge his force, in order to prepare for chase, combat, or flight; and likewise, allowing us on land to look inside the fortresses, billets, and defences of the enemy from some prominence, although far way, or also in open campaign to see and to distinguish in detail, to our very great advantage, all his movements and preparations; besides many other benefits, clearly manifest to all judicious persons.[9]

Unsurprisingly, given this particular emphasis on covert observation, the telescope can be found referred to as an *espia de vidro* (glass spy) in a Spanish work of the 1630s, eventually passing into wider usage in English as "spyglass."[10]

The telescope's promised effects soon began to be felt on the battlefield. By the first half of the seventeenth century, figures such as English admiral William Monson were openly bemoaning the spread of telescopic devices for undermining the effectiveness of maneuvers against the enemy.[11] Navies most readily embraced the telescope because the wide expanses of the seas offered unparalleled open vistas and high contrasts, which suited the effective use of even the most rudimentary devices. The telescope's growing importance for both navigation and offensive operations eventually made it the chief symbol of naval command, as evident by its omnipresence in military portraiture well into the twentieth century. Its adoption by terrestrial forces was more piecemeal until sufficient progress had been made in magnification and field of view, as well as the mitigation of various optical aberrations. However, in due course, the telescope became indispensable to both advance reconnaissance and battlefield command. Martin van Creveld observes that in the eighteenth century, the device "permitted Frederick the Great to establish his headquarters at a fixed location overlooking the battlefield rather than having to rush around it, as was the practice of Gustavus Adolphus a century before."[12]

Initially reserved for scouts and military commanders, telescopic vision diffused throughout the ranks of armed forces as weapon systems' range and accuracy markedly increased during the late nineteenth century. Before midcentury, firearms were predominantly equipped with smoothbore barrels and used loose-fitting spherical balls for ammunition. A common infantry weapon like the musket could not exceed ranges above 75 meters with any accuracy and lost lethal velocity beyond 150 meters.[13] With the adoption of barrel rifling and the conoidal bullet, sharpshooting distances increased considerably. In the case of artillery, only objects visible with the naked eye could be targeted as late as the American Civil War. However, by the onset of World War I, accurate ranges of 20 kilometers or more could be reached. During the Second Boer War, the president of the British Optical Society addressed attendees at the 1900 meeting about these revolutionary changes in gun technology and the unprecedented demand for optical aids they induced:

> The present war has in many respects been a revelation to military men, and in none perhaps more so than in the great increase of the distance at which the opposing troops come into actual conflict. With the modern rifle and the modern field gun, not to mention the guns of position which have been so freely used in Natal, it is scarcely an exaggeration to say that the conflict begins before the opposing forces can clearly see one another. It is therefore of very much greater importance now than it has ever been before, that optical appliances should be largely supplied as mere aids to vision, for use during an action, as well as for scouting purposes.[14]

The meeting's attendees also discussed the transition of chemical propellants from black powder to nitrocellulose, which not only contributed to the increase in the range of firearms and artillery but also greatly reduced the visibility of their location in the absence of conspicuous smoke. This innovation in explosives technology promoted the increasingly effective practices of sniping and concealment of artillery positions, as the British army found at considerable expense in South Africa. In his firsthand account of the Boer War, Arthur Conan Doyle observed, "With black powder it was useless to hide the gun as its smoke must betray it. With smokeless powder the guns are so invisible that it was only by the detection with powerful glasses of the dust from the trail on the recoil that the officers were ever able to localise the guns against which they were fighting."[15]

Notwithstanding developments on land during this period, the navy remained the primary source of demand for optical devices through advances in gunnery control and the need for periscopes to equip the emerging submarine fleets. The resounding triumph by the Japanese navy over its Russian counterpart at the Battle of Tsushima in 1905 was attributed in no small part to the performance of a Japanese gunnery aided by superior equipment and training in optical range-finding. In the intense naval arms race between Britain and Germany, investment into optical equipment soared, dwarfing the corresponding sums being spent for ground forces.[16]

The systematic deployment of optical aids for land operations was brought about by the stalemate of entrenched positions during World War I. Telescopic vision became highly prized, with each side intensely scrutinizing the other to determine the adversary's defensive configurations, movements, and intentions. Ensuring a large-scale supply of so-called optical munitions soon acquired paramount strategic importance for France and Great Britain, who both had allowed the peacetime industry to be dominated by German producers. In November 1914, the British Optical Society was informed that 6 or 7 percent of an army that now counted around a million men (and soon to rapidly rise) would require binoculars, and 1 percent would need optical range finders.[17] For prismatic binoculars alone, the volume of orders in the first year of the war represented a twentyfold increase on that of the previous year, and it was soon evident that domestic firms would be unable to meet that immediate demand.[18] Although supply problems were never entirely resolved, the newly created Ministry of Munitions established an Optical Munitions and Glassware Department and tasked it with coordinating production and procurement, a belated but real recognition of the optics industry's importance to the war effort. Before the conflict, France had similarly imported 70 percent of its optical instruments from Germany. Through a comprehensive mobilization of its own domestic industry, the monthly output of optical glass grew to 12 tons and the concurrent production of prismatic binoculars increased from 1,500 to 13,000 units over the course of the war.[19] Although the optical munitions industries in both countries would markedly scale back after the Great War, military reliance on its products would only deepen further. Noting his country's critical dependence on American and British suppliers at the outbreak of World War II, Australian journalist H. C. McKay submitted that the glass lens was nothing less than "the giant's eye of modern mechanized warfare, without which armies would be practically blind."[20]

Optical devices are thus vital for surveying fields of operations and observing the positions, movements, and actions of friendly and enemy forces at distances far in excess of the naked eye. Militaries naturally place great value on such intelligence for the conduct of strategic and tactical maneuvers and the general planning of defensive and offensive operations. Yet the fundamental entanglement of vision and optics within sociotechnical formations of military violence can only truly be understood through the specific exercise of weapons targeting—or as Paul Virilio contends, "For men at war, the function of the weapon is the function of the eye."[21] The full elucidation of Virilio's observation requires a systematic analysis of how the sense perception of a discrete entity's presence and motion in Euclidean space is translated into the effective delivery of force against it. This investigation reveals an ever more extensive and sophisticated entrainment of the faculty of vision within the lethal logistics of the war machine, correspondingly detaching it from its corporeal and phenomenal dwellings and rearticulating it within new machinic architectures.

The Four Orders of Targeting

In charting the mobilization of vision for increasingly remote projections of military force, we can distinguish four orders of targeting: aiming, ranging, tracking, and guiding. A few words about the choice of the term "order" are first necessary, however. It should be primarily understood here as the arrangement or disposition of constituent elements and their relations that characterizes these processes of targeting. But "order" also connotes the authoritative command or disciplinary instruction entailed by the corresponding marshaling of bodies and machines. Finally, if there is also an historical and functional logic to the sequence in which each schema is presented, they do not thereby supplant one another. Rather, each order enfolds the previous one within its operation. The ranging of distance to a target necessitates taking visual aim at it. Tracking a mobile entity in space and attempting to predict its future trajectory requires a steady stream of range measurements. Finally, guiding a projectile to its intended destination involves continuously tracking their relative positions. The four orders are therefore all present in the contemporary battlespace, variously instantiated within specific military assemblages.

The first order of targeting corresponds to "the act of taking aim," which Virilio describes as "a geometrification of looking, a way of technically aligning ocular perception along an imaginary axis."[22] With projectile weapons capable of striking adversaries at a distance, the line of sight tends toward an alignment with the line of fire, as evidenced by the sling, bow, or firearm. Where the provision of the projectile's motive energy is entirely delegated to the weapon itself, the required skills from its human operator are primarily those of visual acuity and accurate aim. Accordingly, sheer physical prowess becomes less important than the nervous system efficiently linking the visual cortex to the appendage tasked with triggering the delivery of deadly force, often reduced to a single digit in today's war machine. The line of sight thus becomes a lethal perpendicular ray, focused through the frame of iron sights, crosshairs, and other reticles that realize the alignment of eye, weapon, and target.

As firearms' range and accuracy progressively increased, telescopic lenses permitting the optical magnification of distant objects provided ever greater assistance to the act of aiming, above all for specialist marksmen. The first systematic use of telescopic sights occurred in World War I. German forces led the way, with 20,000 sights reportedly in service by the end of 1914. According to one British officer's account, exposure of any part of the body for any longer than three seconds would invariably draw German sniper fire. With telescopically equipped rifles

FIGURE 10. *British and Australian troops with observation periscope and periscope rifle, Gallipoli, 1915. Copyright Imperial War Museums (Q 13427).*

having accurate ranges of 300 meters, direct hits frequently ensued.[23] For this reason, so-called trench periscopes came into widespread use. Whether for stand-alone observation or firing from cover when affixed to a rifle, these devices enabled a continued optical alignment of the gaze with the weapon while simultaneously effecting a physical displacement of the eye through the refraction of light via prisms or mirrors (Figure 10). This role of telescopically enhanced vision on the battlefield only increased with the subsequent gains in the lethal ranges attained by sharpshooters. By World War II, targets at distances of 600 meters could be commonly hit. Today, for skilled marksmen with favorable conditions, reliably accurate fire can be achieved for ranges of 1,200 meters, and although exceedingly rare, lethal shots from over 2,000 meters away have been reported. No longer the sole preserve of snipers and other specialist weapon operators, telescopic sights have become a standard accessory issued with infantry rifles for leading militaries.[24]

Although frequently mediated by lenses and other reflecting or refracting surfaces, the first order of targeting is fundamentally constituted by an optical alignment of eye and target along a line of sight that a weapon system's trajectory strives to approximate. Of course, any physical projectile, subject to the forces of gravity and atmospheric friction, inevitably follows a parabolic trajectory that can never be strictly aligned with a visual ray. While line of sight and line of fire still broadly correspond in the common uses of modern firearms, a shooter has to account for the arcing path of a bullet when firing at a target lying outside of a given weapon's optimal operational range by estimating distance and either deliberately aiming off target or adjusting the optical sights accordingly.[25]

This important relationship between distance and parabolic trajectories brings us to the second order of targeting, which encompasses the visual techniques of measurement that determine the precise range from a weapon to a target. Originating in methods of land surveying, the military practice of visual ranging developed alongside the artillery revolution in early modern Europe.[26] When combined with the science of ballistics, ranging greatly improves the targeting accuracy of any munition that takes a marked parabolic trajectory and where the simple alignment of sight, weapon, and target is therefore inadequate. Exercises in early gunnery were essentially a combination of rule-of-thumb guesswork and trial and error, with any necessary adjustments to an artillery piece's aim made by the observed divergence between the target's position and that of previously missed shots. From the sixteenth century onward, this largely intuitive practice was supplemented by a range of cognitive aids and instruments such as artillery tables, theodolites, military compasses, and gunners' quadrants. Among their various purposes, these devices assisted in the application of the surveying technique of triangulation for determining the distance between two points. This trigonometric technique had been known since antiquity, but it was only rediscovered in Europe during the Renaissance. For any given triangle, it is possible to mathematically deduce from the known values of one of its side's lengths and two of its angles the value of all the remaining sides. Thus, for the purposes of artillery, a target's range can be determined from the measurement of the distance between two surveying positions and of the respective angles formed by the lines of sight from these positions to the target (Figure 11). While there is still a geometric alignment of vision with the imaginary axis through which the surveyor takes aim at the target, this alignment no longer purports to

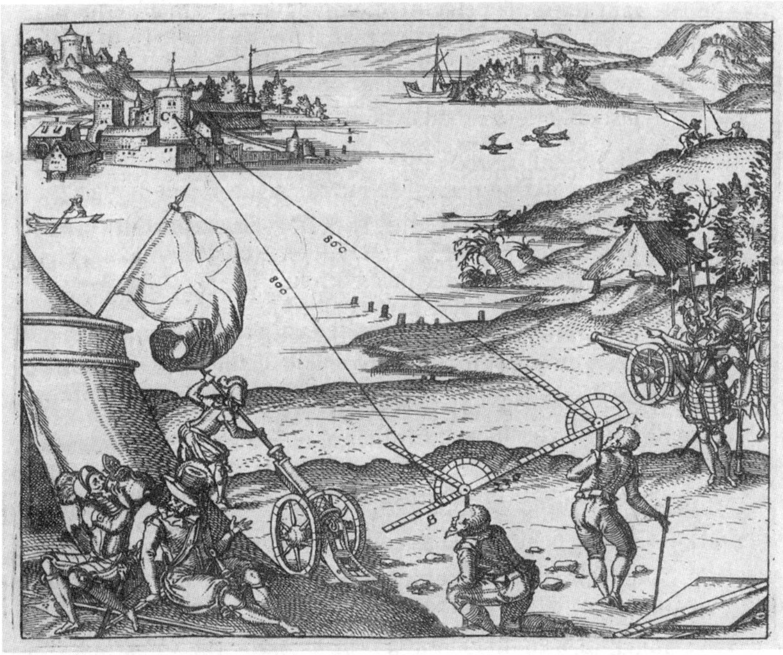

FIGURE 11. *Range-finding through triangulation. Leonhard Zubler,* Novum instrumentum geometricum *(1607). SLUB Dresden / Deutsche Fotothek.*

approximate that of the weapon's line of fire. Instead, the phenomenal unit of visual perception is harnessed into a calculative assemblage of geospatialization through the abstract recombination of distinct lines of sight.

Building on these trigonometric principles, a number of optical devices were eventually designed to permit range-finding in the field through the unified phenomenal frame of a single operator. The simple and reasonably accurate technique of stadiametric range-finding is carried out with a telescopic lens on which is layered a reticle with marks of a known angular spacing. For objects of a known size, such as a soldier or a tank, their distance can be easily determined through the telescopic sight by reference to the reticle's markings. Introduced during World War I for artillery use, sights adapted for stadiametric range-finding continue to be widely used for both their ease of use and inconspicuousness—qualities particularly valued by snipers.

The optical range finder, a more powerful and accurate apparatus, was conceived at the end of the nineteenth century and was thereafter widely adopted by militaries on land and at sea. The range finder represents an ingenious technical condensation of the visual surveying technique. The telescopic device consists of an array of lenses and prisms arranged along an elongated tube, the length of which serves as the perpendicular base of a right-angle triangle to the object being sighted (Figure 12).[27] The two images entering the apparatus from both ends are brought together in either one or two central eyepieces, depending on the chosen operative mechanism. The coincidence range finder makes use of a single eyepiece, with each image respectively occupying the upper and lower halves of the eyepiece display. The operator is tasked with aligning the two separate images of the target being sighted by manually orienting the device's internal optical elements. Once the images coincide, the distance to the target can simply be read off the range finder's instrumentation, with all the necessary trigonometric calculations having been automatically performed. In contrast, the stereoscopic range finder makes use of two eyepieces, one for each image. Here the operator's visual cortex has to merge both images, producing a stereoscopic image that generates a sense of depth. Adjustments are then made to position the target at the same visual distance as reference marks positioned in the eyepieces, with range again displayed on the range finder's instruments.[28] Although the stereoscopic model is, at least on the surface, closest to unassisted human binocular vision, both types of apparatus realize machinic concatenations of discrete lines of sight that demand training an operator's vision in such a way as to perform the device's function of optically determining distance.[29]

In addition to the visual determination of the distance between a weapon and its intended target, the surveying technique of triangulation also permits the sighting of a target to occur at a remove from the weapon's position through relatively straightforward exercises in geospatialization. Such procedures are invaluable to the practice of indirect fire where no direct line of sight from an artillery gun to the intended target exists as a result of impaired visibility, intervening obstacles, or the need for protective cover. In the most basic configuration, surveying takes place from the vantage of a third position where both target and weapon system can be sighted. The surveyor registers the two ensuing measurements on the gridded surface of a plotting board, which then yields the relative positions of one surveyed entity to the other by way

FIGURE 12. *Naval range finder in a target practice exercise, 1913. Harris & Ewing Collection, Prints & Photographs Division, Library of Congress (LC-H261-3193).*

of simple trigonometric calculations (Figure 13). In effect, the plotting board allows for the creation of a crude map, further evidence of the intertwined histories of cartography and targeting that will be examined at greater length in chapter 4.

The third order of targeting encompasses techniques that use the continuous tracking of a mobile target to predict its future position and schedule an encounter with a lethal projectile. Just as advances in gunnery were delivering marked improvements in the ability to target specific points in space, the growing mobility of targets in the twentieth century posed a new problem, the resolution of which would require the considerably more exigent task of targeting points in space-time. Indeed, beyond a certain range, the motion of a nonstationary target requires the weapon system's anticipation of the target's position at the moment when the fired projectile will reach it. The corresponding practice of "leading" the target by firing ahead of its present position is one that has long been carried out intuitively by archers and riflemen. Yet in the last century, the formidable increases in the range of weapon systems and the mobility of their potential targets have necessitated

FIGURE 13. *Forward observation post, Royal Canadian Artillery, featuring range finder, binoculars, plotting board, radio, and field telephone (azimuth telescope in the background), Italy, 1943. Lieut. Alex M. Stirton, Canada. Department of National Defence, Library and Archives Canada, PA-177156.*

techniques of high-speed computation to add prediction to the existing military assemblages of perception and localization.

The inception of this predictive capability can be traced back to the first decades of the twentieth century and the efforts by major naval powers to improve the accuracy of long-range fire against other ships. The necessary calculations were fiendishly complex, requiring information on the target's range, speed, and course as well as the firing ship's own motion (its speed and course but also its roll, pitch, and yaw), along with other variables relevant to directing artillery fire.[30] Naval gun control consequently saw the functions of perception, localization, and gun laying being disaggregated and redistributed across the ship now constituted as an integrated weapon system. Through the use of long-base range finders and precision telescopes, officers in elevated director towers sighted the position of targets along with the locations where missed shots fell. Firing orders were issued to a room below deck where gunnery officers plotted data points, calculated estimations of future positions, and, on the basis of available artillery tables, passed on the relevant firing solutions to the personnel in charge of the ship's guns.[31] Initially carried out almost entirely manually, the fire-control chain underwent progressive mechanization and automation of its functions. The inconclusive Battle of Jutland in 1916 had exposed the weak performance of the supposedly leading fire control systems of the British navy, with only 3 percent of shots hitting their intended German targets. However, the battle simultaneously highlighted the superior operation of the only British ship equipped with a mechanized calculating system.[32] In the following decades, incremental automation of the targeting process moved forward until information gathered from the sighting of enemy units could be directly transferred electrically to the plotting room where, factoring in the ship's own motion sensors, predictive firing solutions would be automatically calculated and transmitted to servomechanically controlled and gyroscopically stabilized guns. At the very heart of these integrated systems for fire control was the range keeper, an electromechanical computing device that calculated the current target bearing and generated predictions as to its future position for translation into firing instructions.[33]

However, through the interwar years, then with ratcheted intensity during World War II, the primary driver of the automation of targeting within space-time was the so-called antiaircraft problem. On land and sea, the task of fire control was immensely complicated by the appearance of airplanes, which could move at high speed and in three dimensions. The additional spatial dimension proved to be the lesser issue because the analytical problem of targeting was not fundamentally different from the two-dimensional case mastered through existing fire-control techniques. The central challenge lay instead in the drastic reduction in the time available for processing and communicating effective firing instructions from tracking observations. The enhancement of antiaircraft defenses thus became a major priority throughout World

War II, with the Office of Scientific Research and Development (OSRD), the agency set up to coordinate scientific research for the American war effort, leading much of the work. An array of OSRD-funded projects dedicated to various aspects of the antiaircraft problem furthered the systemic integration of the constitutive elements of perception (supported by optical range finders and tracking telescopes at first, thereafter progressively displaced by the new electromagnetic technology of radar), computation (executed by devices known as range keepers, predictors, gun directors, and eventually simply as computers), and gun laying (increasingly performed by electrohydraulic remote control). At the outset of the war, as many as fourteen antiaircraft personnel were required to perform the tasks of observation and tracking, with the information being manually relayed to the gun controllers. The implementation of automatic control and the electrical relaying of information would markedly reduce the reliance on manpower while dramatically improving the speed and reliability of execution.[34]

These innovations did not render the human component insignificant, however. The human operator remained essential to these systems, but its corporeal embedding into the machine deepened. Gyro gunsights, such as the U.S. Navy's Mark 14, required the operator to take visual aim at a target aircraft, keeping the sight's reticle locked on the plane as it moved through the air, while the device calculated the necessary lead and adjusted the gun's actual aim (Figure 14). Its manufacturer promoted the new sight in 1941 by claiming that it "broadens the mental powers of the gunner, frees him from tasks requiring judgment, and enables him to devote his entire attention to the accurate 'tracking' of enemy aircraft."[35] The gunner, no longer having to intuit the trajectory of the target, could become a dedicated servomechanical vision machine. The continuous tracking by the gunner's eye provided the vital informational input for the sight computer's aiming of the gun. The same principle is at work in the tachometric bombsights deployed during World War II for the conduct of aerial bombing. The famous Norden bombsight involved the bombardier adjusting the settings on the bombsight's state-of-the-art mechanical computer until the sight's visual crosshairs remained stationary as the bomber approached the target. With momentary control over the aircraft's speed and course handed over to it, the bombsight then automatically triggered the bomb release on the basis of a complex real-time calculation involving the aircraft's speed, altitude, and windage.[36]

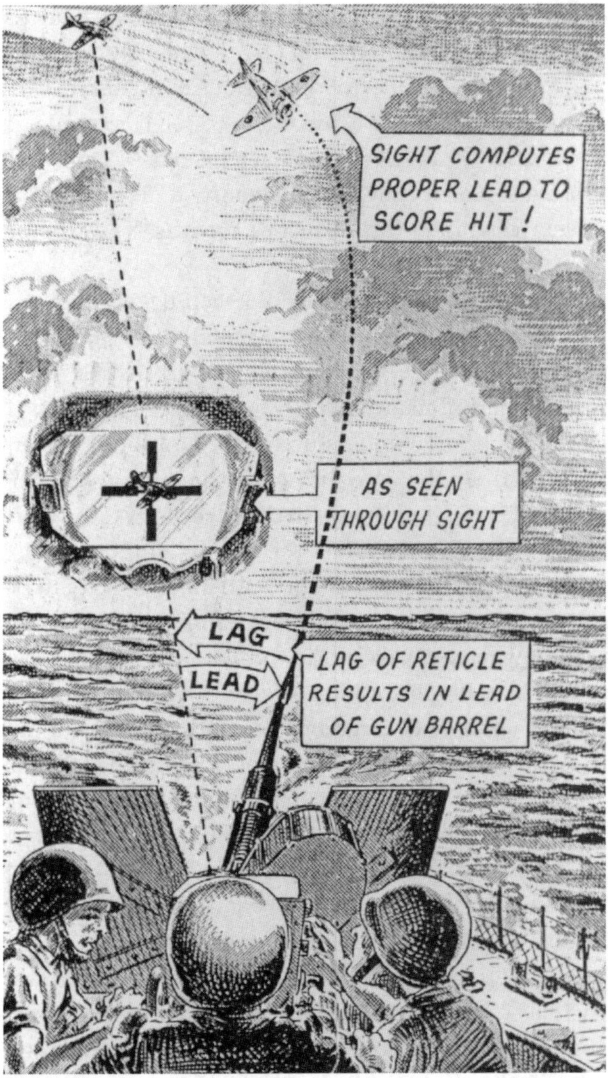

FIGURE 14. *Principle of operation of the Sperry-Draper Mark 14 gunsight. The operator on the right adjusts the sight's range setting on the basis of his estimation of the target's distance. Image scan courtesy of David Mindell.*

For Zabet Patterson, these targeting apparatuses are "specifically dedicated to augmenting, informing, and enframing the soldier's process of seeing" and "directly shape the actions of which he is capable" through "a process in which the human body is reeducated by the machine to act according to a new paradigm of visuality."[37] The act of seeing becomes synonymous with the rapid scanning of the visual field for a movement or object of interest, followed by a focusing and locking of the gaze on an identified target as it is continuously tracked across the field. The eye's sustained fixation on the target generates the data stream necessary to automatically derive predictions of the target's future behavior and direct deadly force against it—"to see is to model is to comprehend is to destroy."[38] Yet as the eye itself becomes simultaneously disciplined into a visual regime of calculability and control, the optical sight effectively (en)frames the eye just as much as the would-be target.

This regime of militarized perception would seep deeply into the postwar scientific culture through its computational reinterpretation of the human organism. Having served during the war in the Royal Air Force's Signals branch, influential British neuropsychologist R. L. Gregory put forward a cybernetic understanding of how humans perceive movement. He described the "two movement signalling systems" of the "image/retina" and "eye/head" in which the edge of the retina acts as "an early-warning device, used to rotate the eyes to aim the sophisticated object-recognition part of the system on to objects likely to be friend or foe." Using terms explicitly derived from gunnery, "where similar considerations apply when guns are aimed from the moving platform of a ship," Gregory articulated a conception of vision framed in terms of a dynamic process of tracking and targeting.[39] As Jordan Crandall underlines, this cybernetic paradigm of vision and targeting has become the template for the detection, processing, and codification of moving phenomena. Whether the phenomenon in question is "a stock price, a biological function, an enemy, [or] a consumer good," its tracking and the resulting generation of future predictions promise to afford a "real time perceptual agency" and the gain of a strategic advantage in the relevant "competitive theater." Tracking thus stands as the "dominant perceptual activity in a computerized culture where looking has come to mean calculating rather than visualizing in the traditional sense and where seeing is infused with the logics of tactics and maneuver."[40]

Yet for all the technical advances made during World War II, the antiaircraft problem could not be solved on an entirely ballistic basis. Accurate prediction of a highly mobile target's future position at a precise moment in time was simply found to be impossible to achieve reliably, particularly at the supersonic speeds of travel available soon after the war. Several sources of uncertainty conspire to undermine attempts at this kind of extrapolation. For one, an aircraft's trajectory is rarely

FIGURE 15. *Director tower for the antiaircraft High Angle Control System (HACS) on HMS* King George V *during World War II. The crew involves a control officer (rear of the unit, with binoculars), layer, trainer, and range taker (front, face hidden). Copyright Imperial War Museums (A 3883).*

smooth, as both the pilot and an array of mechanical and aerodynamical factors introduce unintended irregularities. Second, machines and their operators also inject significant measurement errors (or noise) in the tracking data, a problem that became particularly acute with the introduction of the electromagnetic technology of radar. Finally, even if a successful technical system could satisfactorily mitigate these uncertainties, enemy pilots would inevitably learn to take evasive action to frustrate any prediction of their flight course.[41]

This impasse gave rise to a fourth order of targeting. Guidance systems, which permit the adjustment of a projectile's trajectory in midflight, relocate elements of the targeting process into the projectile itself. While a general estimation of the ideal ballistic path before the projectile's release remains desirable, a guided munition has the additional capacity to steer itself toward its target and correct for any deviations. Whether entirely self-contained within the munition or spatially distributed across discrete entities, fully automated or requiring manual input, all varieties of guidance systems are essentially composed of the same core functional components. In the terms of art, a sensor acts as the munition's "eyes" in establishing and maintaining a "line of sight" to the target. A control system thereupon establishes the relative geometry between munition and target, contrasts it with the desired geometry, and generates the steering commands necessary for their convergence. Effectors then translate these commands into the required mechanical adjustments, be they the sole orientation of aerodynamic control surfaces (as with guided bombs) or their combination with thrust control (as with self-propelled missiles). By extending the process of tracking until the very point of impact, guidance dramatically enhances the accuracy of targeting over large distances far beyond anything possible with simple ballistic fire control.

Pursued by all sides throughout World War II, the technical challenge of implementing weapon guidance found a ruthless but effective solution in the Japanese kamikaze pilot. As a postwar report by the American National Defense Research Committee recognized, "The simplest method of obtaining target discrimination is through its recognition by intelligence," and "the Japanese suicide missiles employed exactly this technique, using human organisms to guide the missiles." Although the report determined that "economy as well as considerations of humanity" recommended against the use of humans, the recruitment of "lower organisms" into the process of targeting was nonetheless pursued in one of the most singular research programs of the war.[42] Directed by psychologist B. F. Skinner, who would subsequently find fame for his behaviorist work, Project Pigeon proposed to entrust the guidance of missiles to the eponymous bird.

This peculiar guidance system's underlying principle was simple enough, as described by Skinner in his subsequent account. The animal was to be tightly ensconced in the missile's nose, where "an image of the target was projected on a translucent screen as in a camera obscura. The pigeon, held near the screen, was reinforced for pecking at the image on the screen. The guiding signal was to be picked up from the point of contact of screen and beak."[43] Through the fastidious conditioning of its reflexes, the bird could be made to ignore any distractions occasioned by loud noises, vibrations, and acceleration so as to respond only to the required visual stimuli and issue the necessary corrections of any deviations by the missile as it journeyed to its intended destina-

FIGURE 16. *Three-pigeon missile guidance system. Courtesy of the B. F. Skinner Foundation.*

tion. Skinner's research showed that, after a careful process of selection, those pigeons that proved most responsive to the operant conditioning could, through exposure to relevant aerial photographs, be successfully trained to follow a variety of targets on land and at sea. They would disregard other objects in the visual field and consistently prioritize a single target when several were visible. In order to hedge against any aberrant pecking by a single animal, the design of the homing pigeon system evolved to house three birds, the sum of their individual behaviors "democratically" determining the guidance instructions to be conveyed to the missile (Figure 16).[44] The OSRD saw enough promise in Project Pigeon to award it $25,000 in June 1943. While his program was canceled the following year, Skinner would later maintain that it had less to do with the actual performance of the guidance system than the challenge of convincing his superiors to take seriously his avowedly "crackpot idea."[45] Skinner's contention seems vindicated by the Naval Research Laboratory's subsequent decision to revive the project in 1948 under the denomination of ORCON (for "organic control"). The program was supported until 1953, by which time the superiority of electronic guidance systems had become incontrovertibly established.

Despite its lack of fruition, the entire endeavor should be considered as more than just an eccentric footnote in the history of guidance technology. Following on the previously outlined regimentations of bodily vision, Project Pigeon presents us with the stark marshaling of organic sense perception and its attendant network of reflexes into a control mechanism for the extension of the targeting circuit into the projectile itself. Conditioned to ignore all other possible stimuli and harnessed to exclude any other freedom of movement, the pigeon becomes a cybernetic control unit that tirelessly processes a flow of visual information into a corresponding set of corrective actions, steering its vehicle toward the intended target and thereby making itself the unwitting agent of its own annihilation. Or, as Skinner put it to his funders, "We have used pigeons, not because the pigeon is an intelligent bird, but because it is a practical one and can be made into a machine, from all practical points of view."[46]

Seeking an alternative to the wholesale assimilation of the living organism into the guided weapon, the U.S. military recruited the technology of television into military assemblages of remote perception and control. As an electro-optical system for the transmission and reception of moving images at a distance, the invention of television fired the military imagination almost immediately. The radio transmission of live images promised a definitive untethering of the human eye from its corporeal shell, allowing it to view remote locations and events in real time without the requirement of a direct line of sight. After a demonstration of the new technology in 1933, one U.S. admiral noted that the military could now "scout the enemy with television equipment in a plane, direct the fire of our gunners and make great advances in aerial mapping."[47] Above all, the tantalizing prospect of conjoining television with radio control would bring perception and distant action under the unitary phenomenal frame of a single operator.

In 1934, Radio Corporation of America (RCA) engineer Vladimir Zworykin proposed a scheme for a "flying torpedo with an electric eye," in effect a glide bomb fitted with television and remote control. As Zworykin explained, "After it has been released the torpedo can be guided to its target with short-wave radio control, the operator being able to see the target through the 'eye' of the torpedo as it approaches."[48]

Under this arrangement, the human eye becomes coterminous with that of the missile, hunting down its target until the latter fills the entire field of vision, the very instant at which both perception and its object are simultaneously obliterated. Although it received no immediate funding, Zworykin's proposal outlined the first concrete scheme for the use of television in missile guidance, which would gain new momentum with the outbreak of World War II.

During the actual conflict, television ended up playing only a marginal role. RCA delivered over four thousand television systems to the U.S. military and developed an image orthicon camera tube with a much improved sensitivity to light compared to its predecessors.[49] However, using television for remote guidance in the manner that Zworykin had imagined proved to be a considerable challenge given the necessary miniaturization of the equipment and the difficulties posed by radio signal interference. Despite the substantial expenditure of resources, these obstacles could not be overcome in time to make a meaningful contribution to the war. Only a few experimental weapon systems were ever deployed in the field. The Navy did assemble a television-guided pilotless aircraft to be used as a flying bomb against naval targets, envisioning the production of a thousand of these so-called assault drones. However, persistent technical problems and the military brass's loss of interest meant only a fifth of the original order was fulfilled, and operational deployment was limited to a few sporadic attacks against the Japanese. In the European theater, the Air Force's secret program Operation Aphrodite sought to remotely guide war-weary B-17 bombers laden with explosives into selected targets. A pilot and flight engineer would be responsible for getting the bomber airborne and setting it on course before bailing out and handing over control to the crew of a second aircraft tasked with leading the bomber to its target through a televised feed. The Air Force attempted over a dozen missions, but every one of them ended in failure, and the program had to be abandoned.[50] With Germany's work on television guidance for its Hs 293 glide bomb proving no more effectual, the American military's postwar assessment was that "no television-missile system was successful as World War II closed."[51]

Subsequent efforts proved more fruitful. The successful deployment of the American television-guided bomb Walleye in the Vietnam War led to a wider adoption of successor weapons such as the Maverick

AGM-65A/B missile and the GBU-15 glide bomb. With all these systems, the operators did not need to guide the missile all the way to its target. Instead, they could opt to use the onboard camera to visually lock on to a target before handing control over to automatic guidance. Today, live video feeds are essential to the operation of remotely piloted vehicles, supporting the targeting process of unmanned hunter-killer aircraft such as the MQ-1 Predator or MQ-9 Reaper. Yet television is just one of many sensor technologies that have been integrated into post–World War II guidance systems and more broadly within the perceptual arsenal of the modern war machine.

Just as the four orders of aiming, ranging, tracking, and guiding are copresent in contemporary military assemblages, each of their functions is liable to be fulfilled through any number of investments of the perceptual field. For both historical and expository reasons, we have focused almost exclusively on the human eye's innate perception of the narrow spectrum of visible light and the optical enhancements that have been made to this faculty. Yet the martial gaze's mobilization of perception extends far beyond this original optical realm and its inherent limitations. Norbert Wiener pointed out in 1950 that "every instrument in the repertory of the scientific-instrument maker is a possible sense organ."[52] It is to the military recruitment of these organs that we now turn in surveying the enlisting of the wider electromagnetic spectrum, the marshaling of sound frequencies, and even the regimenting of individual photons into concentrated beams of light.

Beyond the Visible Spectrum

In the latter half of the nineteenth century, scientists discovered that visible light registered by the human eye comprised only a narrow sliver of the frequencies (or wavelengths) that make up a much broader electromagnetic field. The reach of our perceptual nets has dramatically expanded through the progressive annexation of this full spectrum of radiant energy, with profound ramifications for both the conduct and experience of war. While all electromagnetic frequencies hold military potential, the lower frequencies of infrared, microwaves, and radio waves are the portions of the spectrum most relevant to the present-day operation of the martial gaze.[53] The harnessing of acoustic phenomena is similarly instructive to the processes by which sensorial perception is rationalized, mechanized, and integrated into wider targeting assemblages.

Throughout military history, nocturnal maneuvers and combat have been attractive to commanders for the natural protection that darkness affords, allowing for undetected movement and the potentially decisive gift of surprise. However, operations in darkness also present considerable risks, including spatial disorientation, communication breakdown, loss of command, and friendly fire. Sources of illumination such as torches, flares, or flashlights provide greater visibility but are equally likely to draw unwanted attention. Perceptual prosthetics developed to compensate for the innate limitations of the human eye in darkness therefore amount to one of the most consequential extensions of military perception, engendering the possibility of twenty-four-hour battlefields and bestowing a vital advantage to any armed force that can claim to "own the night."[54]

Although William Herschel first discovered evidence of infrared radiation in 1800, the practical means to render it visible to the human eye were not available until the first half of the twentieth century. The first infrared photographic film appeared just before the Great War but remained an immature military technology until World War II, when it came into use to counter camouflage efforts directed against aerial reconnaissance. The invention of the photoelectric image tube would prove to be the more consequential technological breakthrough in the long run, however. An outcome of early television research during the interwar period, the photoelectric image tube is capable of capturing infrared and other nonvisible frequencies of light and converting them into an electronic signal that can be fed back into a live visual display. During World War II, the first night-vision devices to make use of such image tubes included the American M2 Sniperscope (Figure 17), the Soviet Iskra gunsight, and the German rifle scope Zielgerät 1229 (code name Vampir). All these contraptions saw limited deployment, hampered by their restricted range and the bulkiness of their short-lived battery packs. The requirement for an active infrared searchlight to illuminate their target area, thereby rendering an operation readily detectable by enemy forces equipped with similar equipment, was a further drawback common to all these early systems. The Vietnam War saw the fielding of the first generation of entirely passive night-vision

FIGURE 17. *Australian sniper with the American M2 Sniperscope during the Korean War. Photograph by Claude Rudolph Holzheimer. Australian War Memorial (AWM 147849).*

devices, which dispensed with the need for infrared illumination. These apparatuses benefited from improved image intensifiers that electronically amplified and projected photons from existing ambient light onto a monochromatic display in the eyepiece.[55] The Starlight scope notably provided an image sharp enough to enable shooting out to 400 meters. According to one source, the scope was responsible for 15 percent of night operation kills despite its limited availability, considerable optical distortion, and operational requirements of a clear sky and half-moon conditions.[56] Subsequent generations of image intensifiers have overcome these restrictions, offering substantially improved image resolution and a fiftyfold increase in light amplification. An array of scopes and goggles now routinely mediates the vision of infantry, special forces, tank crews, and aviators during increasingly frequent nocturnal operations.

A parallel line of development to the amplification of light is the detection and visualization of the electromagnetic radiation emitted by hot bodies. After piecemeal experimentation during World War II, the

U.S. military began equipping missile guidance systems with thermal-infrared detectors in the 1950s, beginning with the Falcon and Sidewinder air-to-air missiles.[57] By identifying and tracking an enemy aircraft's infrared signature, chiefly the heat emitted by their engines and exhausts but also by the friction of airflow on the vehicle's body, such missiles can relentlessly pursue their targets without the need for any further human direction. Starting in the late 1960s, forward-looking infrared (FLIR) cameras that convert thermal radiation into video output have equipped airborne platforms for assistance in navigation, reconnaissance, and targeting. As FLIR systems have become more portable and energy efficient, they have diffused throughout the various arms of the military. Although offering a lower resolution than image intensifiers, thermal infrared cameras can range to greater distances and penetrate not only the darkest of nights but also smoke, haze, and fog.

After incremental advances over several decades, the technologies of image intensification and thermal imaging were dramatically showcased in large-scale night maneuvers during the 1991 Gulf War. General Barry McCaffrey, responsible for leading the famed left-hook flanking attack into Iraqi territory, concluded that the U.S. military's "night vision capability provided the single greatest mismatch of the war."[58] Although the widespread diffusion of night-vision devices has partially eroded this imbalance, the execution of complex tactical schemes under cover of darkness is still considered to be one of the key sources of global military superiority. Militaries are pursuing ongoing research and development of supporting technologies, alongside the study of the various perceptual, cognitive, and physiological "human factors" that these prostheses augment.[59] The latest generation of portable devices, such as the U.S. Army's Enhanced Night Vision Goggles (ENVG), combines image intensification with thermal imaging on a single screen display and enables the distribution of sensor imagery through digitization and wireless networks. Future technologies may include graphene infrared sensors so thin they can be integrated into contact lenses, thereby affording wearers a much wider field of vision than permitted by existing optical devices.[60]

While vision has been the preeminent martial sense and certainly the first to undergo systematic rationalization and prosthetic augmentation, other sensory modalities have been similarly mobilized. The sense of hearing has always played an important role in war, with sound being audible to human ears across a 360-degree range, supplementing and directing the much narrower field of vision. Just as with the powers of sight, the technical amplification of hearing beyond rudimentary, if occasionally ingenious, instruments and its integration into wider assemblages of location and targeting hinged on the constitution of a science of acoustic phenomena. From its modern foundations laid by Galileo during the Renaissance, the science of acoustics advanced rapidly in the eighteenth and nineteenth centuries, establishing a theoretical understanding of the properties of sound waves and obtaining empirical measurements of the speed at which sound travels through various media.[61]

Once again, World War I would provide the opportunity for the systematic military deployment of this scientific knowledge through an assortment of practical schemes and devices. Harnessing human physiology into acoustic apparatuses of varying sophistication, binaural listening apparatuses relied on large horns so an operator could hear a phase difference between sound waves reaching each ear. By adjusting the position of the horns until a sound reached both ears simultaneously, the operator could ascertain the general direction of the sound's source (Figure 18). While binaural listening devices proved somewhat successful for ships and aircraft, background noise was too great an obstacle to enable the reliable location of land artillery.

Instead, the technique of sound ranging proved to be the more effective method for acoustically locating enemy guns in a war dominated by indirect fire.[62] Sound ranging necessitated the precise measurement of time intervals between the registration of the sound of artillery fire at discrete stations. Using established scientific knowledge of sound waves' behavior, operators could then use these intervals to deduce the position of the battery. While the theoretical principles of sound ranging were generally understood at the outset of the war, its dependable technical implementation in the field was far from realized as each side devised different methods. German sound ranging relied on trained human listeners equipped with stopwatches to record artillery detonations, while the Allies worked on instrumental recording techniques that entailed a greater objectification of perception and ultimately proved far more accurate.[63]

The Allies' sound-ranging sections typically involved six microphone stations situated at carefully surveyed points along a wide base behind

FIGURE 18. *Future physics Nobel Prize laureate Jean Baptiste Perrin (left) helps operate a binaural acoustic aircraft detector in France during World War I. AIP Emilio Segrè Visual Archives, Gift of Leonard Loeb.*

two advance observation posts, all of which were electrically wired to a central station. When the forward observers reported artillery fire, a galvanometer in the central station would transcribe onto photographic film the electrical impulses transmitted by the microphone stations as they registered the low-frequency sound waves emanating from the enemy guns. With a sensitivity of measurement on the order of a hundredth of a second, the film markings—with the necessary adjustments made for the effects of wind and air temperature on traveling sound—determined the arcs interpreters drew, with lengths of string, from the indicated positions of the respective microphone stations on an area map. The gun being ranged lay at the intersection of these arcs and could thereafter be submitted to counterbattery fire. By the end of the war, the system could locate within 50 yards (45 meters) artillery positions that lay at ranges of 10,000 yards (9,000 meters) in no more than a few minutes.[64] Moreover, in picking up the sound of a shell explosion, the microphones enabled the sound-ranging units to determine the location of its impact. Sound rangers could further derive not only the shell's time of flight and range but also the type and caliber of the originating gun.[65] Finally, armies could also apply the same techniques to track and adjust their own artillery fire. The formidable operational achievements of sound ranging, based on just a theoretical capability at the conflict's outset, was the outcome of considerable resources and scientific expertise, an effort that amounts for Peter Chasseaud to no less than the Great War's own "Manhattan project."[66]

Technologies for acoustic location continued their development after the war. Britain erected a number of monumental concave concrete structures known as sound mirrors in coastal areas to serve as static dishes for picking up the engine sound of incoming aircraft and supplying vital advance warning to the country's air defenses. Although Britain conducted lengthy trials and made plans for a nationwide network of sound mirrors, it soon became apparent that increasing aircraft speeds perilously shrank the window of time available to respond, given the system's optimal detection range of 20 miles (32 kilometers). The scheme was eventually abandoned in the mid-1930s in favor of radar technology's superior range and accuracy. While radar would eventually supplant acoustic techniques almost everywhere, the technology's relative infancy at the onset of World War II meant that belligerents still deployed a wide range of acoustic locators, particularly for the direction of searchlights to locate enemy aircraft. However, these locators were generally found to be unsatisfactory. Artillery sound ranging proved to be a more resilient technique and continued to perform an essential function within counterbattery fire throughout the conflict.

The most enduring applications for sound location can be found underwater, where other modes of perception such as optics or radar are severely hamstrung. In 1490, Leonardo da Vinci observed that the sound of distant ships could be heard at the end of a long submerged tube. But this simple method could determine neither the ship's direction nor range. Not until the late nineteenth century, with the imperative to safeguard the growing maritime traffic from accidental collisions, did underwater acoustics receive the sustained attention necessary for the elaboration of more sophisticated detection devices. Already stepped up in the aftermath of the *Titanic* disaster in 1912, research efforts redoubled with the outbreak of World War I and the grave threat to Allied shipping presented by Germany's furtive U-boats. For John Shiga, the maritime dimension of the conflict saw an "intensive and widespread process of sonic weaponization" that resulted in the enduring "enactment of a new model of subsurface space as a combat environment" in which "isolating a target in the acoustic field and separating it from ocean noise became the necessary first step toward destroying it."[67]

This militarization of the subaquatic domain gave rise to a cluster of underwater acoustic techniques eventually referred to as sonar (for "sound navigation and ranging"). These techniques can be grouped into two essential varieties of passive and active listening. The former rests on similar principles to sound ranging on land, using arrays of hydrophones (underwater microphones) to detect the underwater sounds emitted by enemy vessels. Active sonar, on the other hand, involves emitting a pulse of sound (or "ping") and listening for any echo produced by its reflection off an object in its path, with the intervening delay providing a measurement of distance. While active sonar typically possesses superior range and accuracy, it has the marked drawback of potentially revealing both its activity and position to an adversary's passive sonar. With any breach of silence potentially fateful in the deadly games of hide-and-seek played out in the stygian depths of the planet's oceans, submarine commanders are instructed to use their active sonar sparingly.

Throughout the Cold War, the United States built up an extensive passive sonar network by laying hydrophone arrays on the continental shelves of the Atlantic and Pacific oceans. Faced with a possible rerun of World War II's Battle of the Atlantic, the U.S. Navy deployed a system for long-range undersea surveillance that would "turn the ocean itself against the Soviet Navy" by using the aquatic expanse's "own characteristics to identify submarine activity."[68] Tapping into the so-called deep sound channel, a horizontal layer of oceanic water whose temperature and water pressure enable low-frequency sound waves to travel thousands of kilometers before dissipating, the Sound Surveillance System (SOSUS) could triangulate the position of any Soviet submarine approaching the North American coastline. Whereas operators of other sonar systems usually required finely attuned hearing,[69] SOSUS graphically represented acoustic signals by tracing an electrostatic stylus on heat-sensitive carbon paper. Analysts subsequently interpreted the spectrograms, which required the acute ability for discerning the "subtle nuances in sound signals via intensity, color, shape, and shade that often made the difference between seeing a school of fish or a submarine."[70] In order to recognize the signature of a target's presence amid the subaquatic noise, these self-styled "ocean technicians" had to learn to hear with their eyes, a task for which color-blind analysts were particularly suited as a result of their heightened sensitivity to fine shades of black-and-white.

Today acoustic location is still an indispensable technique in the aquatic realm, where no genuine alternative mode of perception is available, and it continues to play a more discreet yet significant role on land. Advances in signal processing, achieved largely through the development of underwater sonar, enable contemporary acoustic systems to scan for specific target signatures defined by their sonic characteristics of frequency, amplitude, and harmonic relationships. Current devices can detect, identify, and locate ground-based or aerial targets at ranges of several kilometers. A recent area of application is localizing the origin of small-arms fire that presents too low a cross section for radar detection.[71] Although radar remains more powerful and versatile for larger targets, the undetectable character of passive listening grants sound location one substantial advantage over existing radar technology.

Operating on analogous principles to active sonar, radar (from "radio detection and ranging") emits pulses of electromagnetic waves—be they radio waves or microwaves—and receives reflections from any objects these waves encounter. The distance, position, motion, and identity of these objects are then derived from the recorded delay and characteristic features of the returning signals.[72] Ranging over distances far greater than telescopic vision and unaffected by variations in luminosity or cloud cover, the ability of radar to detect and locate enemy units as well as precisely scan and recognize terrain features has made it an indispensable equipment of modern militaries, extending their perceptual reach far beyond that of visible light or infrared.

The British, building on their early acoustic antiaircraft defense scheme, were the first to deploy a comprehensive operational radar system. Under the moniker of Chain Home, they erected a network of early warning radar stations across the southern and eastern coastlines of the British Isles in anticipation of German bombing raids. Individual stations passed on details of incoming and outgoing aircraft to so-called filter rooms. Staff would then collate these reports and plot the latest positions and altitudes of both friendly and enemy forces on large gridded map tables. Drawing on this synthetic view of the battlespace, senior officers could issue orders for the coordination of fighter interception and antiaircraft artillery. Although reliant on a relatively crude radar technology, Chain Home conferred a vital advantage to the Royal Air Force's Fighter Command during the Battle of Britain, thwarting German invasion plans despite the Luftwaffe's clear numerical superiority.

Radar went on to play a crucial role during the remainder of the war, deployed by Allied forces on land, air, and sea to locate enemy forces, direct artillery fire, support aerial bombing, and assist navigation. Led by American scientific and engineering efforts after 1940, the Allies achieved dramatic improvements in the range and precision of devices, simultaneously deepening radar's integration into an array of military systems.[73] By the end of the conflict, American researchers had designed over 150 different radar devices and produced close to a million individual units.[74] In terms of targeting, the radar technology may not have proved more accurate than optical sighting in directing guns or dropping bombs, but it was invaluable for operations at night or in cloudy weather. Robert Buderi concludes that, notwithstanding the dramatic unveiling of nuclear weapons at its conclusion, it was radar that truly won the war because "no single technology was more versatile, or had a more pervasive reach or a more devastating effect on the enemy."[75]

After the war, the urgent need to assemble air defense networks that could detect an incoming nuclear attack, by jet aircraft and later ballistic missiles, drove the subsequent technological development of radar. New antenna designs and increasingly sophisticated signal processing further extended the capabilities and practical applications of radar. These capabilities presently include general battlefield surveillance, air traffic control, navigation, high-resolution imaging, terrain mapping, meteorology, missile guidance, and the tracking of artillery and mortar rounds for both fire control and the localization of enemy batteries. Supplementing and often displacing previous optical and acoustic techniques, radar grants modern militaries a veritable sixth sense that leaves contemporary battlespaces awash with microwave and radio-wave frequencies imperceptible to the human organisms that still populate them.

The Weaponization of Light Itself

Originating in postwar research into radar, the discovery of laser (for "light amplification by stimulated emission of radiation") opened up yet another sensorial domain for the previously discussed orders of targeting to harness. Through the controlled release of photons from the excitation of atoms and the shifts in electrons' energy states, laser devices radiate tight directional beams of light of narrowly restricted electromagnetic wavelengths. Depending on the specific material medium being stimulated and the corresponding wavelengths of light emitted, these beams take on different properties such as visibility and color, duration, range, and power. After the assembly of the first working laser in 1960, the technology found a plethora of applications, including telecommunications, engineering, manufacturing, nuclear physics, medicine, and consumer electronics. In the military arena, the laser has been recruited to fulfill all of the targeting functions of detection, ranging, tracking, and guidance. In addition, and most significantly, laser technology makes possible directed-energy weapons with singular properties.

As with the use of sound and radio waves for range-finding, it is possible to precisely measure the time needed for a laser pulse to be reflected off a target. Already deployed in the field by various countries in the late 1960s, laser range finders have almost entirely supplanted their optical counterparts through their superior accuracy, convenience, and operational range. With rapidly firing devices permitting the continuous tracking of mobile targets, laser range finders are a major component of contemporary fire control systems. Furthermore, laser range-finding has given rise to lidar (for "light detection and ranging"), a remote sensing technology that measures the range to every point in a given field of view and generates detailed three-dimensional models of the scenes surveyed with a resolution superior to that afforded by radar.

In the manner of previous perceptual technologies, laser has been widely integrated into the available suite of weapon guidance systems. A typical deployment involves a target designator directing a coded laser beam at a target, allowing a munition's laser target seeker (generally an infrared sensor) to lock on to the beam's reflections and steer a trajectory to the intended destination.[76] Among the first such systems to enter service, the Paveway laser-guided bomb was introduced in Southeast Asia in 1968. Most famously, it was credited with inflicting critical damage on the Thanh Hóa Bridge in 1972 after the structure had survived hundreds of prior aerial attacks. By the end of the war, 28,000 laser-guided bombs had been dropped, and according to one estimate, just under half of them achieved direct hits on their targets, compared to a mere 5 percent success rate for unguided munitions.[77] Unsurprisingly, these reported levels of accuracy drove the further adoption of laser guidance. With the advent of microchip technology permitting the miniaturization of components, laser guidance became a major enabler of the new vision of aerial precision bombing showcased during the 1991 Persian Gulf War. While guided weapons represented a mere 7 percent of the total munitions used, media coverage of the war focused on the telegenic footage of aerial precision weapons, whose leading star was the laser-guided bomb, which has since played a prominent role in all major theaters involving U.S. forces, including Yugoslavia, Afghanistan, and Iraq. Laser guidance does have technical limitations, however. It requires a direct line of sight from the target designator to the target that is liable to be severely impaired by unfavorable atmospheric conditions such as cloud cover, water vapor, and dust or smoke particles.

Although laser has been primarily adopted to support the targeting of bombs and missiles, it is now increasingly equipping smaller ballistic weapons. Laser sights have long been available as a visual aid to the aiming of firearms, particularly in low-light conditions, by emitting a beam or spot that is visible to either the naked eye or a night-vision device. Recently, more sophisticated integrations of laser into firearms have emerged, adapting technology previously reserved to larger weapon systems. In 2013, Texas-based company TrackingPoint introduced a laser-enabled "precision-guided firearm" that claims to dramatically increase first round hit probability at ranges of 1,000 yards

(900 meters) and beyond without the requirement for advanced marksmanship training.[78] The weapon's video scope tags a target by locking a laser range finder onto it, providing an instantaneous measurement of distance and continuous tracking of movement. The onboard computer thereupon calculates the ballistic solution, accounting for such factors as bullet drop, wind, temperature, humidity, and firearm incline. The weapon's operator only has to align the scope's crosshairs with the target tag, and the ballistics computer automatically fires the gun at the opportune moment. By removing the skills and experience normally indispensable for long distance sharpshooting, TrackingPoint proudly proclaims its product to be "democratizing accuracy," edging closer to making sensorial acquisition the sole task demanded of a firearm operator.[79] The U.S. military is already looking ahead to the next generation of technology where laser guidance will be incorporated into the firearm's projectile itself. In 2012, Sandia National Laboratories, a subsidiary of Lockheed Martin, announced the successful testing of a small-caliber bullet capable of steering itself toward a laser-designated target located 2,000 meters away.[80] A similar program, Extreme Accuracy Tasked Ordnance (EXACTO), is also under way under the aegis of the Pentagon's Defense Advanced Research Projects Agency (DARPA).[81]

Yet for all these successful applications of laser, it is its potential as a directed-energy weapon, with all its echoes of the science fiction trope of the death ray, that has most mesmerized military minds from the technology's very inception.[82] In 1962, the head of Army Ordnance Missile Command opined that "laser may be the biggest breakthrough in the weapons area since the atomic bomb."[83] While ambitions for an operational weapon that would revolutionize war fighting in the manner of its nuclear counterpart have to date been frustrated, significant progress toward that goal has nonetheless been made.

The military has already weaponized relatively low-energy lasers for the purpose of blinding light-sensitive receptors, be they electro-optical (particularly sensors operating in the infrared range) or biological—that is, the human eye. A certain irony can be found in the eye's special sensitivity to light, the source of its preeminent role in the support of weaponry, simultaneously rendering it uniquely vulnerable to the invisible rays of weaponized light. After a campaign by the Red Cross and concerned governments, a United Nations ban on "blinding laser weapons" was agreed in 1995, requiring of signatory states that they refrain from using lasers to "cause permanent blindness to unenhanced vision" resulting from retinal burns and intraocular bleeding.[84] The terms of the ban still permit the infliction of temporary blindness and pursuit of attacks against optical equipment (including those that might thereupon cause permanent blindness to their users).[85] A range of laser dazzlers designed to disrupt infrared sensors and assail the human eye are thus in service today.

For the military, however, the real prize is a laser weapon capable of destroying targets such as missiles or aircraft through critical thermal damage, a task that requires much more powerful beams and still faces substantial technical obstacles. The military appeal of a practical high-energy laser weapon is considerable.[86] Unaffected by gravity, laser beams move in a straight line at the speed of light, reaching any target virtually instantaneously. Because it does not have the parabolic trajectory of a ballistic projectile or the need to anticipate the future position of a moving target, a laser weapon greatly simplifies the problem of fire control, increasing the probability of a direct hit. In addition, the narrow focus of a laser beam promises greater target discrimination and reduced collateral damage. Laser beams can also be invisible (at least to the naked eye), making their detection and determination of point of origin more difficult to the enemy.[87] Finally, a laser weapon dispenses with the need for ammunition and its corresponding logistical supply chain, relying solely on a reliable, albeit abundant, power source for its operation. For all these reasons, a high-energy laser appears in principle to be an ideal weapon that could confer a potentially decisive military advantage.

With the theoretical utility of high-powered beams recognized even before the creation of the first laser device, the pursuit of weaponization was coeval with the technology's practical development. In the fraught context of the Cold War, the prospect of shooting down incoming nuclear warheads exerted a particular attraction on the military. In 1962, the Air Force chief of staff, General Curtis LeMay, publicly predicted that "beam directed energy weapons would be able to transmit energy across space with the speed of light and bring about the technological disarmament of nuclear weapons" and warned of the need to get ahead of the Soviets.[88] By 1978, the U.S. Department of Defense had devoted over $1 billion to researching high-energy lasers without having produced anything remotely approaching a useful weapon.[89] While

FIGURE 19. *Airborne laser turret, 2004. Lockheed Martin.*

researchers could engineer laser beams powerful enough to destroy targets in test conditions, the devices' size and energy requirements alone made them impractical.

In the 1980s, the Reagan administration launched a renewed drive under the Strategic Defense Initiative, in particular through a research program dubbed Excalibur that proposed to develop an X-ray laser powered by a nuclear explosion. In what turned out to be a wildly optimistic assessment, Edward Teller, one of the project's chief scientists and already renowned as "the father of the hydrogen bomb," suggested in 1984 that "a single X-ray laser module the size of an executive desk which applied this technology could potentially shoot down the entire Soviet land-based missile force."[90] The lack of tangible progress and the demise of the Strategic Defense Initiative with the end of the Cold War did not end the ambition of fielding laser missile defenses. In 1996, the U.S. Air Force and Boeing began collaborating on the YAL-1 Airborne Laser, which involved mounting a high-energy laser onto an aircraft for the purpose of destroying ballistic missiles in the early boost phase after their launch (Figure 19).[91] Although the $5 billion program conducted a few successful tests against dummy missiles, the Pentagon eventually canceled it in 2012, having failed to establish the practicality of an operational deployment.[92] Undeterred by this setback, the Missile Defense Agency is now seeking to fit the next generation of high-energy lasers onto UAVs.[93] A plethora of other laser weapon systems can be added to these high-profile anti–ballistic missile schemes, with devices conceived and trialed over the years for use against aerial vehicles, rockets, artillery and mortar rounds, mines, and other ground targets.[94] However, most of these systems have only attained limited field deployments, and no high-energy laser weapon can yet claim to have significantly affected the conduct of war.

While dashed optimism has been the norm for the past half century of laser weaponry research, faith in the future viability of such devices and their military advantages appears undiminished. Indeed, the proliferation of government-funded and commercial programs around the world suggests that a number of systems may well enter service in the coming decades, even though their efficiency is likely to be curtailed by laser's continuing sensitivity to atmospheric conditions and potential countermeasures that can materially absorb or reflect laser radiation. Yet the drive toward laser weaponry is certain to endure, so completely does it seem to fulfill the convergence of perception with targeting we have traced here. Already capable of performing all the sensorial tasks required to support the targeting of kinetic force, laser's ultimate promise is to weaponize light itself, marshaling photons into a lethal beam coincident with the visual ray. The line of sight will then truly have become coterminous with the pure, cold line of abolition—one that extends, perhaps, to the annihilation of the very eye that gave rise to it.

CHAPTER 3

Imaging

The camera seems to me, next to unassisted and weaponless consciousness, the central instrument of our time.

—JAMES AGEE, *Let Us Now Praise Famous Men* (1941)

If advances in sensorial capabilities have enabled military perception to range farther and wider than ever, then the full extent of their applications has rested no less on the means devised to reliably capture and circulate visual representations of the battlespace. For Martin Jay, modern visual culture is thus grounded in two main strands of technical development: "the extension of the range and power of our ocular apparatus" on the one hand and "the improvement of our ability to disseminate the results in visually accessible ways" on the other.[1] Having examined the first strand in the previous chapter, we now consider the second's role in the conduct of modern warfare. Indeed, armed conflict has been one of the central drivers in the technical evolution of visual representation through the last century. Friedrich Kittler underlines that "the strategic interest in faster information—the supervising and directing of one's own troops, the monitoring and surveillance of the enemy troops, and above all the supervising and directing of one's response to enemy actions, which should be as immediate as possible—crucially accelerated the explosive rise of optical media."[2] The image, as the surface on which visual information is inscribed and displayed, is common to all the successive incarnations of optical media. Yet the technical infrastructure and fields of application of the image have simultaneously undergone extensive transformations in the past two hundred years, as this chapter will chart with a particular eye to their military ramifications.

We begin with the advent of photography as "the mechanization of the perspective painting and of the arrested eye" and an examination of its revolutionary effects on the consciousness of perception.[3] As William Mitchell emphasizes, the camera has commonly been seen as usurping its organic counterpart: "It is supereye—a perceptual prosthesis that can stop action better than the human eye, resolve finer detail, remorselessly attend to the subtlest distinctions of intensity, and not leave unregistered anything in the field of its gaze."[4] Able to mercilessly fix and capture anything in its aim, the mechanical eye occupies a privileged position within the epistemological arsenal of modern techniques of power, intersecting with the long history of targeting at multiple junctures. The camera instantaneously registers a scene's spatial dimensions within its perspectival images, a feature eagerly seized on by its photogrammetrist interpreters. As it became airborne, photography promptly acquired a central role within military assemblages of targeting, surveying the battlefield and providing vital intelligence throughout the two world wars. The Cold War subsequently launched the supereye into orbit, leaving no stretch of the planet's expanse immune from its unrelenting scrutiny. Concurrent with its ascent into space, the photographic image underwent a radical mutation in the second half of the twentieth century. It became assimilated within the generalized paradigm of electronically processable digital information, a material transformation impelled in no small part by military imperatives. Indeed, the rise of the digital image accompanies that of a new regime of machinic vision, where machines learn to see autonomously and humans are increasingly immersed within synthetic visual environments that entirely overlay their field of perception.

First Exposure

As we have already seen, the Renaissance discovery of linear perspective consisted of an ingenious mathematical technique for the visual representation of physical space from a discrete position that rigorously registers the optical size-to-distance ratios of objects. The invention of photography is quite simply the culmination of painstaking efforts toward its mechanization. For Joel Snyder, it would be a mistake to view the photographic process as "scientific corroboration of the schemata or rules invented by painters to make realistic pictures." On the contrary, photographic cameras represent for Snyder "the incorporation of those schemata into a tool designed and built, with great difficulty and over a long period of time, to aid painters and draughtsmen in the production of certain kinds of pictures."[5] While the knowledge that the passage of light through a pinhole forms an image goes back to antiquity, interest in its use for the construction of realistic pictures did not manifest itself until the diffusion of the principles of linear perspective. Nor were post-Renaissance painters absorbed with manually replicating the appearance of pinhole images in their art but rather with making "a machine that produced an image like the ones they painted."[6] One consequently has to wait for the first half of the sixteenth century to find the first explicit discussions of the camera obscura. The camera obscura, a darkened room in which an aperture was made into one of its sides, allowed for the projection of an (inverted) image onto a blank surface, an outline of which could be traced. Innovators soon added polished lenses to sharpen the focus of images, and over the course of the next century, portable devices that could assist artists in the production of perspectival images became available. For Kittler, the camera obscura "combines for the first time the optical transmission of information with the optical storage of information; the former is already fully automatic, whereas the latter is still manual," and remained so until the discovery of the necessary photochemistry.[7]

French inventor Nicéphore Niépce is generally credited with creating the first photographs in the 1820s. Through a process he dubbed heliography, a permanent impression of light was made on metal plates coated with photosensitive material. The technique required lengthy periods of exposure that essentially restricted its use to the reproduction of still images under controlled light conditions. His collaboration with compatriot Louis Daguerre, which Niépce did not live to see through, would lead to the former's triumphant presentation of the daguerreotype to the world in 1839. The daguerreotype reduced the necessary length of exposure to minutes; subsequent innovations would soon cut this down to mere seconds and less, allowing for the photographic arrest of motion.

The invention of photography immediately sparked huge public interest, provoking animated discussion among the scientists, intellectuals, and artists of the time, and a prodigious production of images followed in the decades after its unveiling. The new imaging technology generated widespread conversations around the nature of perception and the purposes of artistic representation, and it left its stamp on the culture of the nineteenth century as much as the novel. Photography soon became endowed with a singular "reality effect," as many saw in it a more reliable bearer of visual truth than the naked human eye.[8] The images mechanically produced by it were accordingly imbued with an incontrovertible objectivity seized on by scientists and artists alike.[9] Émile Zola, a contemporary writer renowned for his meticulous realistic portrayal of social conditions, felt compelled to affirm "you cannot claim to have fully seen something until you have photographed it, revealing an array of details that could not otherwise be even discerned."[10]

The photographic camera amounted to nothing less than a new organ of perception, possessing a sensorial reach that extended beyond the established limits of the human eye. As William Henry Fox Talbot, British photography inventor and rival of Daguerre, put it, "The eye of the camera [sees] plainly where the human eye would find nothing but darkness."[11] Writing almost a century later, Walter Benjamin would find in the photographic techniques of time-lapse and enlargement the means by which to reveal the "optical unconscious,"[12] that "aspect of material reality [. . .] repressed by the physiology of human vision which can be disinterred by the technically enhanced speed of photographic illumination."[13] For Benjamin, the camera was manifestly giving rise to new subjectivities, which would only be accelerated by the advent of the moving image.[14]

World War I veteran Ernst Jünger also became acutely interested in what he called the "revolutionary fact of photography," notably for its prominent role in the bellicose first half of the twentieth century. He associated photography's advent with the emergence of a "second and colder consciousness," of which the human being was as much the object as the subject. In the manner of Benjamin, he observed that the "artificial eye" of the camera "expose[s] spaces otherwise inaccessible to the human eye" and "penetrates fog banks, haze and darkness, even the resistance of matter itself."[15] However, he went on to vividly assert that photography stood properly "outside of the zone of sensitivity" as "an intensive and invulnerable eye" that "records the bullet in mid-flight just as easily as it captures a man at the moment an explosion tears him apart."[16] He concluded that photographic vision was "an expression of [the] peculiarly cruel way of seeing" characteristic of the new beings that were being forged from the fire of the age of machinic war, individuals for whom seeing was itself "an act of assault."[17]

Jünger ultimately saw photography in terms of "a kind of evil eye, a type of magical possession,"[18] rendering the objects of its capture into epistemic submission. To illustrate the particular power of its objectifying gaze, Jünger invoked the city of Mecca as the example of a cultic site that fell into the "colonial sphere" as soon as it could be photographed. Scientist and politician François Arago's 1839 report to the French parliament on the daguerreotype's potential applications had already outlined this photographic mode of domination. Bemoaning the fact that the technology was not available for Napoleon's conquest of Egypt in 1798 and that the "cupidity of Arabs and vandalism of certain travelers" had since deprived the scientific world of emblematic images, Arago insisted on the value of the new device for the photographic recording of hieroglyphics and survey measurement of ancient Egyptian monuments.[19] It is worth underlining here the place that such exhaustive documentation occupied within the nineteenth century's orientalist project of epistemic subjugation in which, in the words of Edward Said, "knowledge was essentially the *making visible* of material."[20] Yet this activity was only one, albeit important, manifestation of a general impetus to submit the world to the glare of the photographic lens as a mode of possession and ordering. Or as Lev Manovich puts it, "Whenever the real had to be captured, identified, classified, and stored, photography was put to work."[21]

From capture to annihilation, the line was apparent early on. Writing a mere twenty years after photography's invention, Oliver Wendell Holmes proclaimed that "form is henceforth divorced from matter. In fact, matter as a visible object is of no great use any longer, except as the mould on which form is shaped. Give us a few negatives of a thing worth seeing, taken from different points of view, and that is all we want of it. Pull it down or burn it up, if you please."[22] Soon enough, a structural homology between the technologies of imaging and destruction would be revealed in the work of one scientific pioneer of serial photography. Impressed by Eadweard Muybridge's photographic studies of animal locomotion, Étienne-Jules Marey sought to capture birds in flight, leading him to conceive a more suitable alternative to his predecessor's static battery of cameras. In the pages of French scientific periodical *La Nature,* he presented his invention of the chronophotographic gun, a camera in the shape of a rifle that "photographs twelve times a second the object that is being aimed at" (Figure 20).[23] Equipped with a rechargeable cylinder gun in which a photosensitive disc rotated to allow for sequential exposure, the camera operated less like a rifle than the Gatling machine gun invented a couple of decades earlier.[24] For Kittler, Marey's chronophotographic gun reveals the historical coincidence of the movie camera and automatic weaponry, noting that shooting film and bullets are both procedures that "focus on and fix objects moving through space."[25]

This same alignment of camera and weaponry would later resurface virtually unchanged in the photographic devices conceived by the military for target practice. With the advent of aerial combat in the first decades of the twentieth century, the effective training of aircraft gunners presented particular challenges because they were expected to aim their weapons at other planes in a highly dynamic environment. With live ammunition out of the question, gun cameras such as the Hythe Mk III provided a realistic yet safe means to train gunners (Figure 21). Modeled on the Lewis aerial machine gun, the Hythe's trigger activated a camera shutter instead of firing bullets, exposing 120mm film of the target with a superimposed reticle registering the accuracy of the gunner's aim.

The vital role of photography in the warring efforts of the last century lies, however, with the function it came to occupy within wider assemblages of military targeting. In his preface to an anthology of photographic images of World War I, Ernst Jünger observes, "A war that is distinguished by the high level of technical precision required to wage it, is bound to leave behind documents more numerous and varied than battles waged in earlier times, less present to consciousness. It is the same intelligence, whose weapons of annihilation can locate the enemy to the exact second and metre, that labours to preserve the great historical event in fine detail."[26] In this view, the conflict's unprecedented photographic documentation was a logical by-product of the belligerents' quest for mechanical precision in the delivery of lethal force. Indeed, the capacity for the mass production and circulation of perspectival imagery, which can be remotely interpreted and geolocated, has made photography indispensable in the arsenal of modern military power. But before turning to the historical record of its keen adoption by the twentieth-century war machine, we must first consider

FIGURE 20. *Marey's chronophotographic gun. Étienne-Jules Marey, "Le Fusil Photographique,"* La Nature, *no. 464 (April 22, 1882).*

FIGURE 21. *Hythe Mk III camera gun, Post Field, Fort Sill, Oklahoma, 1918. National Archives and Records Administration (17340706).*

the photographic image's inherent geometric characteristics and the early metrical efforts they stimulated. Although their ambitions would mostly be realized much later, the nineteenth-century founders of photogrammetry laid much of the essential groundwork for the eventual mechanical autonomization of vision.

Measuring with Photographs

Like the paintings of the Italian Renaissance, a photograph, by virtue of its perspectival construction, is the geometric projection of a three-dimensional scene onto a two-dimensional surface. Or as Harun Farocki puts it, "a photographic image is a cut, a section through the bundles of light rays reflected off objects in a circumscribed space. Photography reproduces the three-dimensional object on a flat plane, based on the laws of projective geometry."[27] The application of the inverse procedure, using the same laws, should in principle allow the recovery of the spatial proportions of an original scene from its photograph. Indeed, this property of the photographic image was recognized from the very beginning. François Arago's original report on the daguerreotype insisted that because such images were "subject in their formation to the rules of geometry," "the exact dimensions" of objects contained within them could be extracted "with the assistance of very little additional data."[28] In further noting the benefit of the new invention to topographers, Arago may well have had in mind the work of his colleague Charles-François Beautemps-Beaupré. Through his late eighteenth-century excursions to southern Australia, Beautemps-Beaupré had shown that it was possible, if onerous, to produce high-quality coastal maps based on perspectival sketches.

Having previously attempted topographic surveys using freehand perspective drawing and the camera lucida (a popular optical device used as a drawing aid in the nineteenth century), military engineer Aimé Laussedat seized on the new invention of the photographic camera in the 1850s. After numerous experiments and the commissioning of bespoke cameras, he was able to produce in 1861 an accurate 200-hectare map of the French village of Buc and its surroundings by reference to four ground photographs. Mindful of the military applications of his discovery, the French colonel forcefully argued for its value to defenders of fortified positions, as it provided panoramas of "dangerous terrain" that an enemy invasion might traverse.[29] Laussedat defined his general approach of iconometry as "the art of recovering the real dimensions of objects on images in which those dimensions are altered by perspective." When combined with the new imaging technology, the approach became known as metrophotography, a term eventually eclipsed in common use by that of photogrammetry.[30]

Drawing on well-established principles of geometry, Laussedat distinguished two techniques through which spatial measurements could be obtained from photographs. The first applied the rules of projective geometry to derive the relative proportions and positions of elements within a single perspectival image. Laussedat considered this technique of interest mainly to architects and builders because it relied on both the relative proximity of elements and regular geometric shapes. He therefore outlined a second approach better suited to landscapes and geographic terrain that necessitated two photographs of a same scene taken from discrete positions of a known distance from each other.[31] Derived from the familiar trigonometric principles used in surveying, this method held the tantalizing promise of revolutionizing the practice of mapmaking. By dramatically reducing the number of required measurements in the field, this method paved the way for the progressive automation of the manual task of plotting a map.[32]

Notwithstanding Laussedat's pioneering contribution, his work attracted scant interest in his home country. Instead, Albrecht Meydenbauer's architectural work in Germany gave photographic surveying its first widespread applications. The Prussians also seemed more alert to its potential military applications, even dispatching a "phototopographic" detachment during the Franco–Prussian war of 1870–71.[33] Yet for all its promise, as well as the solid grounding of its underlying principles, the practice of photogrammetry was frustratingly slow to develop. Decades after its original conception by Laussedat, it remained too cumbersome and imprecise for recruitment into the commonly used array of surveying techniques.[34] As such, photogrammetric applications in the nineteenth century were largely limited to documenting architectural sites[35] or the topographic mapping of mountainous areas where conventional ground surveying was particularly arduous.[36] Early photogrammetrists faced problems such as optical distortions and chromatic aberrations introduced by available lenses and the difficulties inherent in recognizing terrain features on different photographs. The intense skilled labor required to translate photographic images into topographical maps posed a further obstacle; a range of devices that could partially automate this process only appeared in the first decade of the twentieth century. The potential of photogrammetry as an effective means to both facilitate and enhance the process of mapmaking would only be truly realized when the camera took to the sky and captured the land from above. Thus, as Chasseaud points out, "the true significance of these early developments in ground photo-surveying was that they provided the essential theory and equipment for mapping from air photos," and "much of the early apparatus used for plotting from air photos was developed from that originally designed for use in ground survey."[37]

Although Laussedat reportedly experimented with kite photography, aerial photography's origins are generally credited to his compatriot Nadar, who took to the Parisian skies in a balloon in 1858 to record the first known image so made. Nadar's experiments immediately attracted military interest, and the following year he received generous pecuniary offers to carry out aerial photography in support of the French armies in Italy.[38] However, unsuccessful trials prevented its operational deployment in the field, as the technology was deemed too immature as a result of the required lengths of exposure and the bulky apparatus necessary for developing and fixing images. During the American Civil War, the Union Army Balloon Corps did conduct military reconnaissance under the aegis of scientist and aeronaut Thaddeus Lowe, but evidence that aerial photography was successfully conducted remains scant.

Inventors pursued a variety of vehicles for airborne photography over the following decades, including balloons, kites, and even pigeons. In 1891, a German patent was issued for a photographic device inserted within a projectile and propelled into the air by either an artillery gun or a rocket:

> The projectile is fired high in the air in the direction of the object to be photographed, a charge of explosive being then ignited and the parachute apparatus being thereby set free from the projectile. The parachute opens automatically, falling by itself, the photographic apparatus hanging perpendicularly below it, and by a suitable arrangement taking one or more instantaneous photographs of the positions on the earth below, the parachute being then brought back to the point from which it was projected by a line, one end of which is attached to it, the other being retained at the starting point.[39]

Another German engineer, Alfred Maul, picked up on this idea in 1903 with his own camera rocket designed for military reconnaissance, a version of which was tested during the Turkish–Bulgarian war of 1912–13. Ultimately it would fall to the motorized aeroplane to provide the decisive platform for aerial photography. The First World War created the circumstances for these early sporadic trials to be adopted wholesale within the military, with the scale and character of the European conflict supplying the essential spur for rapid advances in both photographic apparatuses and the means of extracting the information contained within their images.

The Verticalized Eye

As the opening maneuvers of 1914 gave way to a protracted war of positions, observation from the sky was soon identified as the most reliable source of intelligence on the topography of enemy territory and disposition of opposing forces. With a network of serpentine trenches extending across 700 kilometers from the Swiss border to the North Sea, military commanders could not attain anything close to a synoptic overview of combat operations without the assistance of the airborne mechanical eye. In a different context that rings no less true here, Walter Benjamin later observed:

> In general, mass movements are more clearly apprehended by the camera than by the eye. A bird's-eye view best captures assemblies of hundreds of thousands. And even when this perspective is no less accessible to the human eye than to the camera, the image formed by the eye cannot be enlarged in the same way as a photograph. This is to say that mass movements, and above all war, are a form of human behavior especially suited to the camera.[40]

Although some militaries initially resisted the systematic adoption of aerial surveillance, all parties eventually engaged in photographic capture on an industrial scale.[41] By the end of the conflict, the Germans, who possessed the highest degree of relevant technical expertise, were capturing four thousand shots daily. The French were even more enthusiastic, if less proficient, developing by then up to ten thousand photographs every day, while the British lagged behind with only one thousand prints.[42] All sides naturally scaled up photographic reconnaissance ahead of major operations. The American Expeditionary Forces initially joined the war in 1917 with the most embryonic of reconnaissance capabilities but was able to take around a hundred thousand pictures in a mere four days of preparation for the Meuse-Argonne offensive in 1918.[43] The military value of aerial reconnaissance then far outweighed any direct influence that the aircraft of the time could exert on the battlespace. Combat ace Manfred von Richthofen, best known as the Red Baron, readily acknowledged that "often a photographic plate is more important than shooting down an enemy machine."[44]

While early observation relied on the vision of the aircrew and the rough sketches they could draw by hand, photography rapidly asserted itself as a more reliable and insightful means of recording aerial views.[45] A 1918 document from the American Air Service intended for the attention of incoming pilots affirms in no uncertain terms the superiority of photographic reconnaissance, extensively listing "the advantages possessed by the lens and the plate over the eye and the memory":

> It covers a given subject in a thorough manner far more perfectly than the eye, which is disturbed by many causes within and without, seeing as it does only one thing at a time, while the lens sees all at once with perfect impartiality. It eliminates any element of the imagination, for the lens is neither dramatic nor poetic and sees things as they are. The lens can cover a territory wider in scope, and in a shorter time, than the eye and the brain and memory could cover even under the most ideal conditions; and its whole maze of detail being grasped and recorded simultaneously.[46]

However, the intelligibility of aerial photographs required precise knowledge of the imaging apparatus's characteristics and position relative to its object. Information on the plate size, lens focal length, altitude, and angle to the ground of a camera were all necessary in order to derive the dimensions and image scale of the area covered in any single shot.

As a "technology of location"[47] inserted within a wider assemblage of targeting, aerial photography furnished images that armies meticulously scrutinized for the presence of enemy forces and related back to the geospatial frameworks used by their artilleries. A 1918 article in an American scientific review reports on how tight this process had become by the end of the war: "Cases are on record where only twenty minutes have elapsed from the time a photographer snapped his camera over the German trenches until his batteries were playing on the spot shown. In that time the airman had returned to his lines, the photograph had been developed and printed, the discovery made, and the batteries given the range and ordered to fire."[48] The specific needs of the military usually privileged certain types of photographic capture. While panoramic and oblique shots were valued for providing an overview of an area's relief, vertical aerial photographs were preferred for locating enemy positions. Aerial photography taken at a perpendicular angle to the ground, which minimized optical distortion, most closely resembled conventional cartographic representations. Through an analytical process called restitution, these images were used to either update existing maps with recent changes in terrain and enemy positions or generate entirely new topographic maps when current ones were deemed insufficiently detailed or accurate.[49] Reconnaissance flights typically took sequences of overlapping images of the ground below, providing comprehensive coverage of an area, which then were stitched together to compose cartographic mosaics (Figure 22). Conversely, existing maps assisted in the recognition, location, and scaling of incoming photographs. Hence, cartography and photographic interpretation were mutually supportive in lending authority to each other throughout the war.

Interpreting aerial photographs for the extraction of intelligence was an exacting task, demanding consummate skills and extensive training from its operatives, particularly as the countermeasures of camouflage and decoying gained in sophistication. Although essential, analysts did not only have to familiarize themselves with the shapes and proportions of objects or with the telltale signs of recent human activity as seen from the air. They also had to internalize the effect of shadows cast by the sun, the distortions introduced by the optical lens or film warping and shrinkage, and the camera's almost inevitable deviation from a strict perpendicular angle to the ground.[50] In other words, photographic interpreters had to learn to see less as they might from the vantage point of an observer in the aircraft than through the medium of the camera itself. This reconfiguration of human vision is most strikingly apparent in the use of stereoscopy, widely deemed to be the most potent instrument in the analyst's arsenal.[51]

From the mid-nineteenth century on, a scientific understanding of binocular vision developed alongside stereoscopic contrivances that presented a slightly different image of a same scene to each eye, engendering an optical illusion of depth in the spectator. Familiar to contemporary audiences as 3-D film projections popular in the early twenty-first century, stereoscopy was present from the very beginnings of photography, and its images delighted and entertained viewers throughout the nineteenth century.[52] Yet stereoscopy took on an entirely instrumental character in the context of aerial photographic interpretation. Viewing closely overlapping images through a purpose-made optical apparatus proved to be invaluable in determining the relief and elevation of terrain. But whereas earlier parlor demonstrations typically presented stereoscopic scenes in order to simulate natural vision, the stereoscopic effect was here deliberately exaggerated, with the distance between each lens's

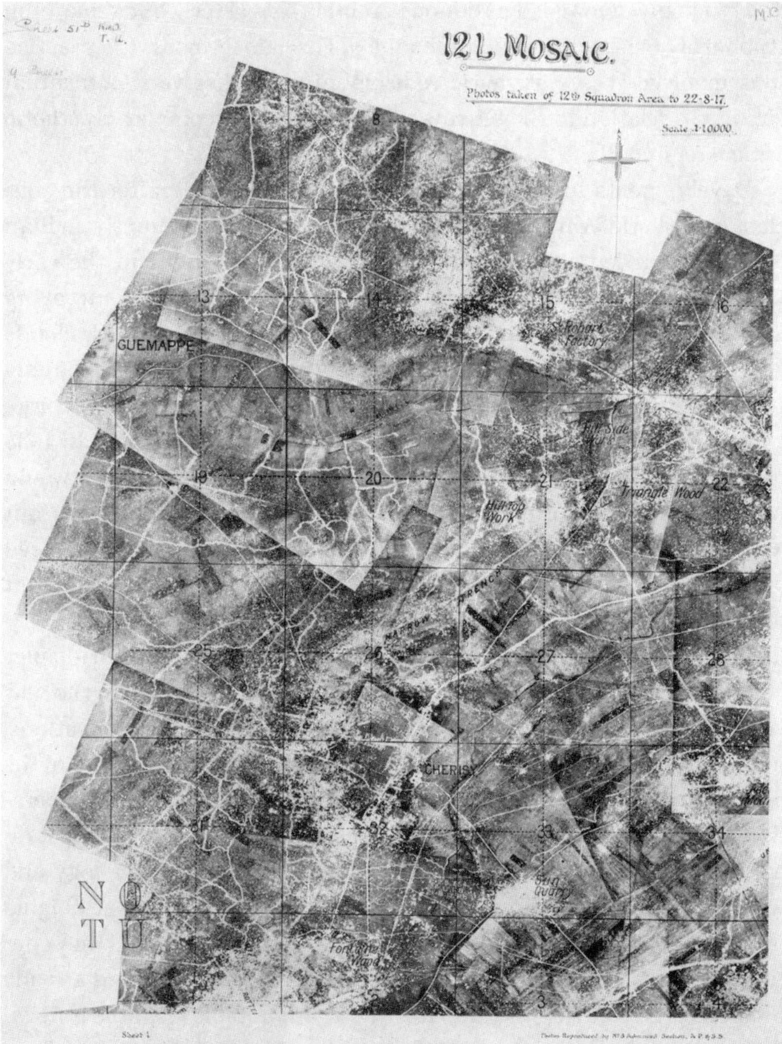

FIGURE 22. *Mosaic map from aerial photographs, 1917. Copyright Imperial War Museums (Q 17254).*

focal point greatly exceeding that between human pupils. This hyperstereoscopy created an effect akin to the analysts peering out of "the surrogate eyes of a virtual colossus," where the lay of the land appeared at a contrast of depth perception inaccessible to the naked eye.[53] However, analysts could not perform this artificial stereopsis without effort, and it usually demanded assiduous practice to master while remaining stubbornly elusive for some individuals. For Saint-Amour, the practice thus involved "the systematic redisciplining of observers' perceptual circuits in relation to the narrow requirements of a complex prosthetic optical regime."[54]

Developments in aerial photogrammetry continued after the conclusion of World War I, if at a reduced pace and predominantly civilian context after demobilization. Significant breakthroughs in the early 1930s definitively established it over conventional ground surveying as the superior means for topographic mapping.[55] As the dark clouds of war gathered over Europe again, airborne photography belatedly regained prominence within rearming militaries. The German Army's commander in chief, Werner von Fritsch, emphatically stated in 1938 that the military organization with the best aerial reconnaissance would win the next war. While the Soviet Union prevailed in the Eastern front with limited capabilities in this area, the "airman's war" would indeed be as much about systematic photographic reconnaissance as about mass aerial bombing.

Because World War I had already established the general principles and operational procedures of aerial reconnaissance, its use in the new conflict is, notwithstanding some piecemeal technological innovations, most significant for its scale and intensity. The positional warfare of the previous conflict gave way to wide-ranging campaigns of mobile operations that needed to be brought under the gaze of the airborne mechanical eye. Through advances in aircraft and photographic technology, and weather permitting, planes could conduct reconnaissance at heights beyond the reach of most antiaircraft ground weapons, even if they continued to remain vulnerable to intercepting aircraft. Film had already replaced plates during the interwar years, but notable wartime innovations included the introduction of infrared photography for camouflage detection and the use of flash bombs for nocturnal reconnaissance. The United States military, whose war effort was highly dependent on aerial bombing, embraced airborne photography with particular verve, generating three million photos every month across the various theaters of war by 1944.[56] One American commander in the Pacific estimated that over 80 percent of his side's wartime intelligence originated in aerial photography.[57]

Air power famously played a key role in the intensification of total war, as so-called strategic bombing of industrial sites and urban centers brought entire societies into the ambit of targeting. Throughout World War I, the nascent practice of aerial bombing had concentrated on the frontline and immediate-rear positions, with only sporadic attempts at striking cities. If the availability of long-range heavy bombers during World War II was essential to this expansion of targeting, then the deliberate planning and execution of strategic bombing relied critically on airborne reconnaissance. Once submitted to the meticulous process of interpretation, aerial photographs provided war planners with the most reliable and up-to-date source of information on an enemy state's industrial and military capacity. Consequently, analysts were no longer only trained to recognize specific military units and sites but were also instructed in the details of complex industrial processes so they could identify any structure that might constitute a high-value target. Once a site of industrial activity had been located, further scrutiny would be applied so as to "estimate annual output, plant capacity (if different from output), the importance of the plant to total national production (percentage of total and, if any, unique production), and potential to convert the industry to other, more directly military purposes."[58] Such estimations would then be revised through a bomb-damage assessment conducted on photographs of a given target taken both before and after the bombing run. In permitting a visualization and geolocation of the material supports of military power, the verticalized photographic image was instrumental to new practices of targeting striving toward the systematic disruption of an enemy conceived as a military-industrial totality.

The Orbital Gaze

The precarious standoff between East and West during the Cold War only further intensified the demand for imagery from above. The United States' and Soviet Union's mutual fear was of an unannounced nuclear onslaught that would lay waste to their societies within the space of a few hours. In the absence of any conceivable countermeasure that might insulate them from atomic devastation, the two sides settled for deterrence through the threat of massive retaliation. Crucially, this so-called balance of terror hinged on a precarious equilibrium between the offensive and defensive capabilities of the two parties. Any failure to maintain this equilibrium, nuclear strategists contended, would only invite the temptation for one state to exploit its temporary advantage and launch a surprise attack on the other. The efficacy of strategic deterrence therefore rested on a continuous assessment of the adversary's arsenal and intentions, from which could be determined the commensurate investment in military assets necessary to counterbalance them. Aerial reconnaissance offered the most dependable means for this task, particularly for the United States, whose efforts to gather human intelligence in the closed society of the Soviet Union were especially arduous.

After the Soviet rejection of President Eisenhower's proposed Open Skies policy, which would have authorized mutual supervised overflights, the United States directed all its efforts toward covert incursions into Soviet airspace. Unmanned balloons flown over Soviet territory under the guise of weather missions met with mixed results and were soon abandoned in favor of a specialist reconnaissance aircraft. Introduced in 1955, the U-2 was designed to fly at 70,000 feet (21,000 meters) to evade radar and antiaircraft defenses while delivering more detailed imagery than World War II aircraft flying at half that altitude. In reality, the Soviets detected the U-2 from the outset, eventually causing its famous downing in 1960, to considerable embarrassment for the U.S. government. Although the incident would motivate the acquisition of the SR-71 Blackbird, a new reconnaissance aircraft with superior maximum speed and cruising altitude, the exacerbated tensions caused by repeated violations of the Soviet Union's sovereign airspace impelled the Americans to search for an alternative means of acquiring photographic intelligence. The solution would be the spaceborne satellite.

Shrouded in a veil of secrecy, the Corona program was instituted to place into orbit reconnaissance satellites that photographed Soviet territory and returned the film to earth via a reentry vehicle. The first test launches began in 1959, with the program experiencing repeated setbacks in its early years. Nevertheless, it soon provided usable images with a ground resolution of 40 feet (12 meters), improving to 6 feet (1.8 meters) or less within barely a decade. As its first major contribution, Corona dispelled the widespread fears of a "missile gap" by bringing incontrovertible evidence that the Soviet Union in fact lagged far behind the United States in terms of nuclear armament. It is in this light that President Johnson's 1967 comments are to be understood:

> We've spent 35 or 40 billion dollars on the space program. If nothing else had come out of it except the knowledge we've gained from space photography, it would be worth ten times what the whole program has cost. Because tonight we know how many missiles the enemy has and, it turned out, our guesses were way off [. . .] Because of satellites, I *know* how many missiles the enemy has.[59]

Satellite reconnaissance became essential to managing the deterrence of full-blown thermonuclear war through the perpetuation of mutual vulnerability. Although by no means the only source of intelligence, satellite imagery provided the most dependable overview of each adversary's capabilities. It would accordingly play a key role in subsequent strategic arms limitation treaties where verifiability of conformity was critical to agreeing to specific restrictions on certain armaments.[60]

Of course, satellite imagery did not just deliver strategic intelligence. It also allowed for the geopositioning of the targets that would be attacked in an eventual nuclear war. By the mid-1960s, the Corona program's photographs were being converted into maps of the Soviet Union more accurate than any in use by Americans for their own country. Onboard cameras captured the positions of surrounding stars, which determined the exact orientation of the satellite and permitted the measurement and subsequent removal of minute optical distortions from mapping images. Satellites could soon locate objects anywhere on the planet at an accuracy of 100 feet (30 meters) or less, thereby supporting the targeting of discrete sites by intercontinental ballistic missiles.[61]

By the time it ended in 1972, the Corona program had overseen 145 launches that had yielded 866,000 frames of photographic film covering a total area just short of 100 million square nautical miles (342 million square kilometers).[62] Having initially stunned the West with the launch of *Sputnik* in 1957, the Soviets responded to the American initiative with the series of Zenit satellites that became operational only a few years after Corona.[63] The American space reconnaissance efforts continued after Corona's retirement with a number of successor programs such as Gambit and Hexagon, all of which would eventually be reclassified as iterations of a single, continuous Keyhole (KH) program, which persists to this day. First launched in 1976, the KH-11 Kennan satellites introduced digital imaging that could be radioed directly to ground stations. These innovations allowed for near real-time electro-optical observation, thereby markedly tightening the reconnaissance cycle. While this capability had been proposed as early as 1951, it could not be realized with the then available television technology and would require the invention of the charge-coupled device (CCD) light sensor.[64] Digital imaging also greatly extended the life of satellites, which were no longer limited by the onboard stock of film and could now operate for months and years rather than mere weeks.

Concurrent with and related to this process of digitization, photography was recast as only one imaging technique within the much wider field of remote sensing. This new perceptual domain encompasses all the means available for the detection and recording of propagated signals across the spectrum of electromagnetic radiation. Multispectral and hyperspectral sensors generate images that cover a wide band of electromagnetic frequencies and from which the spectral signatures of entities that might otherwise evade optical or infrared sensors can be identified. Active sensors direct radio waves, microwaves, and laser beams at target areas and analyze the ensuing reflections. Synthetic aperture radar (SAR) is a particularly valuable technique that produces high-resolution imagery through sophisticated algorithmic processing of radar signals and range measurements. Its principal benefit is the ability to peer through cloud cover, penetrate foliage, and provide imagery regardless of time of day or night at resolutions of a meter or less for military-grade equipment. This capability accounts for the National Reconnaissance Office's claim to "own the night," as trumpeted on the commemorative patch for the launch of a radar-imaging Lacrosse satellite in 2000 (Figure 23).[65]

Optical satellites have nevertheless remained useful, with the latest generation of American KH-12 satellites estimated to have a resolution of 10 centimeters or less.[66] In addition to the United States and Russia, a number of countries—including China, France, Germany, Israel, Italy, Japan, South Korea, India, and Turkey—operate military reconnaissance satellites today.[67] The list of nations with civilian satellites capable of resolutions that would be considered military grade a few decades ago is longer and steadily growing, supplying ever more detailed imagery to the commercial market. Resolutions as high as 25 centimeters are now available on the open international market, forcing a reluctant United States to loosen its restrictions on domestic firms selling satellite imagery.[68]

FIGURE 23. *National Reconnaissance Office patch, 2000.*

For all these impressive technical advances, reconnaissance satellites still have their limitations. Relatively low orbital altitudes of under 1,000 kilometers mean specific areas are only in view of any given satellite during specific moments of their overflight.[69] A competent adversary is therefore quite capable of using these highly predictable orbits to conceal sensitive activities during the passage of known satellites. Multiple satellites do reduce blank spots between overflights, but uninterrupted coverage would require a quantity too costly for even the leading militaries to implement. The United States is nonetheless actively exploring potential means of establishing continuous satellite surveillance of critical areas. One proposal involves the deployment of nanosatellites, small and inexpensive vehicles that could be placed into orbit by the dozen and provide persistent coverage of particular areas for tactical purposes. Once in operation, the system would allow troops to request and receive high-resolution imagery within minutes.[70]

Although a permanent position above a specific point on the globe can be maintained at a geosynchronous orbit of 36,000 kilometers, this distance prevents the acquisition of photographic images at sufficient resolution, which to date has limited satellite reconnaissance to the monitoring of radio frequencies.[71] Attempts to overcome these restrictions are ongoing, however. In 2010, DARPA publicized a project known as MOIRE (Membrane Optic Imager for Real-Time Exploitation), inviting design proposals for the development of a new type of reconnaissance system. MOIRE involves placing into orbit an extremely thin membrane 20 meters in diameter to collect light from the earth and focus it onto a geosynchronous satellite's imaging equipment. With access to an area of 10 million square kilometers, the proposed system would, at any given moment, cover a zone of a 100 square kilometers at a video update rate of at least one frame per second, enabling it to report a tactical ballistic missile launch in near real time.[72]

If satellites permit the surveillance of any spot on the globe without infringing on sovereign airspace, then active military operations still rely heavily on airborne platforms, as these retain the benefits of greater maneuverability and proximity to their targets of observation. In recent years, the marked trend has been a move toward UAVs, among which the RQ-4 Global Hawk presently stands as the most sophisticated surveillance platform. Equipped with remote sensors for optical, infrared, and radar imaging, the Global Hawk can reach similar altitudes to that of the U-2 and can survey up to 100,000 square kilometers during a single day of uninterrupted flight. According to one report, the Global Hawk flew only 3 percent of all aircraft imagery-collection sorties in the 2003 invasion of Iraq but amassed information on 55 percent of "air defense–related time-sensitive targets."[73] Of course, because all military UAVs are necessarily equipped with remote sensing equipment, they are inherently well suited to reconnaissance tasks. Hand-launched miniature drones such as the RQ-11 Raven can provide ground troops with local tactical intelligence, while armed aircraft such as the MQ-1 Predator and MQ-9 Reaper combine the roles of sensor and weapons platforms.[74]

In 2009, the U.S. Air Force unveiled Gorgon Stare, a video-capture technology intended to equip an airborne platform with up to twelve distinct cameras, where each could simultaneously track an individual target within a 4-kilometer radius.[75] The defunded program's specific fate is less interesting here than what its moniker reveals about the imaginary of military surveillance. Just as the mythical creature it drew its name from could turn the objects of her gaze into stone, this all-seeing eye promised to immobilize any entity caught by its ceaseless roaming of the battlespace. A similar inspiration can be found in DARPA's choice of name for their ARGUS-IS (Autonomous Real-Time Ground Ubiquitous Surveillance—Imaging System) program. Argus Panoptes was the name of a hundred-eyed giant who served as a restless watchman in the service of the Greek gods. ARGUS-IS combines 368 optical sensors into a single 1.8-gigapixel camera that enables persistent surveillance over a 6-kilometer-diameter area from an altitude of 6,000 meters. In order to handle the identification and tracking of objects within the vast volumes of imagery generated, artificial intelligence software is entrusted with automatically monitoring every significant movement within its field of view and alerting human analysts to any sensitive activity.[76]

Such visual algorithms are the product of the past decades' cumulative efforts to endow machines with perceptual faculties of discernment that would match (and exceed) our own. The attainment of that goal still faces considerable obstacles. Yet autonomous machine perception is arguably the inexorable logical conclusion of the process of visual rationalization initiated in the Italian quattrocento. For Paul Virilio, the advent of such a "vision machine" and its "instrumental virtual images" is tantamount to the realization of a new "prosthesis of automatic perception that will function as a kind of machinic imaginary from which we will this time be totally excluded."[77]

It Sees!

As the process of imaging is increasingly automated and integrated within wider sociotechnical assemblages, it becomes ever more appropriate to speak, following Harun Farocki, of "operative images." Quite simply, these are images "that do not represent an object, but rather are part of an operation."[78] Any aesthetic quality to these images is entirely incidental because their production is wholly guided by the efficiency with which relevant information can be algorithmically extracted from them. Operative images undoubtedly predate the age of computing, as Jan von Brevern argues in his treatment of nineteenth-century photogrammetry: “Whatever aesthetic qualities contemporaries might have seen in photogrammetric images—they were merely the ‘waste products’ of purely utilitarian aims: measurability, precision, comparability.”[79] Nevertheless, the development of high-speed electronic computation is behind the recent exponential proliferation of these images, most of which are no longer ever viewed by human eyes.[80]

The move to digital electronic imaging denotes a profound and highly consequential shift in the underlying material ontology of the image.[81] Whereas photography registers intensities of light via photosensitive chemicals, electronic image acquisition uses the transduction of radiant energy into patterns of electric current to render images as digitally processable information. Digital imaging sensors are typically composed of arrays of light-sensing diodes, each of which translates the emission of photons from a single pixel (or “picture element”) in the image field into an electric current to constitute a statistical ensemble of discrete numerical values. In effect, the electronic image is a mathematical object before being a visual one because its composition as a statistical set can be expressed through any number of equivalent visualizations. In Edmond Couchot’s words, the digital (or numerical) image

> manifests itself as a matrix of numbers (a table composed of columns and rows) contained in the memory of a computer and capable of being translated through the form of a video or print image. One can henceforth integrally synthesize an image by providing the computer with the matrix of values adequate to each of these points. Inversely, a conventional image, such as a photograph or a drawing, can be analyzed numerically; in that case, the computer decomposes it automatically by the means of a specialized camera and transfers it into its memory in the form of a matrix of numbers.[82]

As an ensemble of discrete numerical values that instruct its display, the digital image can be shaped and manipulated by algorithmic procedures with much greater ease than its analog counterpart. Consequently, the digitization of imaging has considerably advanced the central ambition that already lay behind photogrammetry: “to supplant the human dimension in the measurement process—just as photography had supplanted it in the recording process—to complement extant methods for the photographic recording of a space with a mechanism for the automated derivation of the measurements of a space, to automate the domain of metrology so that it might become sympathetic with the already automated (and instantized) visual domain."[83] A wealth of ingenious optical–mechanical devices dedicated to this end appeared from the early twentieth century onward, including the stereocomparator (1901), autostereograph (1908), autocartograph (1921), and aerocartograph (1926). These early systems assisted greatly in the plotting of maps from photographs, but all continued to rely on intense skilled labor. The alternative approach of analytical plotting via an algebraic method was too computationally onerous for practical applications until the advent of modern calculating machines in the 1950s. Thereafter, plotting via an algebraic method began displacing the older analog approach, laying the groundwork for the contemporary fully digital environment, in which the entire photogrammetric process of image acquisition, derivation of spatial measurements, and representation via a geographic information system (GIS) can be conducted electronically with a high degree of automation.[84] Yet significant as they are in their own right, these developments in photogrammetry are still only one manifestation of vision's digitally enabled autonomization.

Computer vision is the field most evidently associated with this broader trend. Under its umbrella, an array of algorithmic techniques has been devised since the 1960s for the derivation of numerical or symbolic information from images so as to interpret their contents.[85] Feeding into a range of decision-making and control procedures, computer vision is critical to the automation of any task reliant on visual perception and interpretation. Although increasingly replacing human vision in a growing number of activities, none of the presently available computer vision solutions approaches the sophistication and flexibility of their biological model, particularly in an open environment. Such an achievement would hinge on the promise of genuine (or "general") artificial intelligence, the prospect of which is an eminently fraught philosophical question that cannot be resolved here.[86] Nevertheless, the more limited systems of computer vision that have been assembled have indubitably found far-reaching fields of application, not least among the military.

As military sensors and intelligence-gathering activities generate ever-larger volumes of imagery, human operatives are increasingly tasking computer vision software with sifting through reams of data in order to draw out those images most likely to contain items of interest. Imagery obtained through satellite and aerial reconnaissance was among the first to be processed by pattern-recognition and change-detection algorithms, highlighting suspicious structures or alterations on the ground for further consideration.[87] However, the most ambitious applications of computer vision lie in its integration into autonomous systems of weapon navigation and targeting that dispense altogether with human input.

One notable instantiation of computer vision can be found in the DSMAC (Digital Scene Matching Area Correlation), which guides the U.S. Tomahawk cruise missile in its terminal phase of approach to its designated target. DSMAC compares imagery acquired by an onboard optical sensor with stored imagery previously obtained through aerial or satellite reconnaissance. Any measure of displacement between the two images provides the necessary correction signal for the missile's guidance system to home in on its target with an accuracy of a few meters.[88] Attempts to implement scene matching for missile guidance date back to the 1940s with systems that endeavored to compare live televisual input to reference photographs. But it was not until the late 1970s that requisite advances in remote sensing, microelectronics, and precision measurement provided a workable solution. During the 1980s, the untested Tomahawk encountered abundant skepticism throughout the military. After its distinguished performance in the 1991 Gulf War, the missile has since become one of the most prized weapon systems in the American arsenal.[89]

The core principles of such computer vision systems are well established today, although their practical implementation has presented durable obstacles that have only been partially overcome. At its most elementary level, computer vision statistically correlates the intensity values of resolution elements (such as individual pixels in a raster image)[90] within an acquired image to those of preinstalled reference images. If a sufficient degree of similarity for a positive match is found between two images, then the system determines that it has recognized a particular scene or object and submits this information to further decision-making procedures. However, this straightforward method is insufficient for all but the most basic and controlled exercises in computer vision. Imaging an open three-dimensional environment is heavily dependent on such factors as the relative position of the sensor to the captured scene and conditions of illumination and weather, all of which present significant challenges when correlating different images of a single scene. It is therefore generally preferable to identify key features with certain invariant characteristics from the range of discontinuities present in any given image.[91] By linking these features to those of previously stored three-dimensional models and accounting for variations in appearance, computer vision can more accurately match not only particular scenes but also objects and figures within physical space, which can in principle be distinguished, recognized, and located.[92]

First demonstrated in 1963, the algorithmic application of projective geometry to single perspectival images was an important early technique for the automated recognition of three-dimensional solids.[93] Because of the limitations in resolving the visual ambiguities within many images, advancements in machine perception have ultimately relied on a much wider array of sensor inputs and computational techniques for the reconstitution of a scene's three-dimensional properties. Analysis of the perspectival images produced through the encoding of light intensities by a photosensitive sensor have been supplemented by distance measurements directly obtained by ranging instruments such as radar, sonar, or laser. With no general-purpose solution in existence, the exact configuration of computer vision techniques used in any specific application is entirely dependent on the task at hand.

The absence of any general solution makes clear that progress in computer vision has been much more labored and piecemeal than its early pioneers had anticipated, with many of the expected breakthroughs in artificial intelligence still proving elusive. As the Tomahawk demonstrates, machine perception is today an established and reliable enabler of autonomous navigation for guided weapons traveling through known terrain toward designated target destinations.[94] Fully automated acquisition of previously unknown targets, in contrast, presents a considerably tougher challenge than the simple recognition of already identified targets, particularly in complex and cluttered ground environments. While human discernment of targets in these contexts is still markedly superior to that of electronic algorithms, the latter's superior speed and applicability across the full sensorial spectrum of imaging remains highly attractive and continues to draw significant research funding. In 2011, DARPA launched a program called Mind's Eye for advancing the "machine-based visual intelligence" that would arise from "automating the ability to learn generally applicable and generative representations of action between objects in a scene directly from visual inputs, and then reason over those learned representations." Asserting that computer vision has to date mostly focused on the recognition of objects in the world—"the nouns in the description of a scene"—the program's stated goal was to "add the perceptual and cognitive underpinnings for recognizing and reasoning about the verbs in those scenes."[95] By visually identifying certain actions (a man digging up a road, for instance), artificial intelligence could potentially infer the nature of a scene (the planting of an explosive device).

Spurred by the high-profile deployment of UAVs, animated discussion has sprung up around the prospect of future weapon systems equipped with computer vision and synthetic reasoning being granted full autonomy over the decision to use lethal force. In 2013, a U.N. report highlighted major issues surrounding the emergence of "lethal autonomous robotics," calling for national moratoria on their development.[96] Whether or not states heed such a call, persistent limitations in computer vision's ability to reliably acquire new targets or interpret the content of scenes within complex environments still stand in the way.[97] As such, field applications are likely, for the foreseeable future, to be restricted to narrow, time-sensitive scenarios (such as missile defense) or the assistance of human decision-making.[98] The widespread arrival of fully autonomous killer robots may therefore not be imminent. And while our machines are slowly accruing their own independent visual faculties, new synthetic visual environments and informational architectures that cut across the increasingly porous boundaries between the organic and inorganic are concurrently rearticulating human perception.

Machinic Visions

Fateful as it may prove to be, gradual autonomization constitutes only one face of the emergence of "machinic vision" as a "generalized and extended condition of visuality in the era of information machines."[99] Indeed, digital imaging simultaneously enables the presentation of a virtually unlimited array of visualizations to the human senses. Machinic vision, as opposed to mere machine vision, is for John Johnston "not so much a simple seeing with or by means of machines—although it does presuppose this—as it is a decoded seeing, a becoming of perception in relation to machines that necessarily also involves a recoding."[100] The mechanical disembodiment of vision correlates with a biophysical reembodiment, the increasingly complex articulations of which are always further displacing the action of perception from any fixed locus.

The emergence of machinic vision is intimately tied to the intense efforts by the military to grapple with the acute information-processing challenges confronting its operatives in the modern war-fighting environment. In the pursuit of new conveyances of relevant information for its optimal treatment by human cognition, military-funded research has consistently been at the forefront of far-reaching innovations in the field of human–machine interaction. The electronic image here serves as a crucial interface, immersing combatants ever further into synthetic visual environments through which the battlefield is mediated to them.

Over the course of the twentieth century, military planners realized that the proliferation of sensors and the increasing complexity of equipment and tactics placed growing cognitive demands on combatants. This difficulty was most apparent in aerial combat, where airplane and helicopter pilots engaged in a variety of offensive and defensive maneuvers in highly dynamic environments. Cockpit designers confronted "the problem of displaying the increased amount of information [to] be synthesized" and for which the further extension of already crowded instrument panels appeared an inadequate response.[101] In the late 1950s, the heads-up display (HUD) emerged as a first solution. A HUD is a fixed transparent display overlaying the forward line of sight from the front of the cockpit canopy on which live flight or targeting data can be projected.[102] By the 1970s, the miniaturization of displays allowed them to be mounted onto helmets or goggles and positioned immediately ahead of the wearer's eye. Crucially, these head-mounted displays (HMDs) not only allowed visual information to be continuously present within the user's field of view but also responded to the user's movements through head-tracking technology, thereby merging action and perception.[103]

Targeting applications swiftly followed. By linking a weapon system to an HMD so that its aim automatically follows the user's line of sight, a look-and-shoot capability becomes possible.[104] The U.S. Army deployed an early electromechanical realization of this scheme on its AH-1S Cobra attack helicopters in the 1970s, directing the fire of the machine gun turret by reference to the position of a reticle sight mounted on the pilot's helmet. However, the lineage of look-and-shoot weapon systems can be traced all the way back to World War I. A series of patents were then issued to American inventor Albert Pratt for what might today be called an "integrated helmet-mounted aiming device and weapon delivery system."[105] The contraption consisted of a helmet within which was mounted a gun with an affixed sight directly aligned with the wearer's eyes (Figure 24). A tube extending to the user's mouth provided the trigger for the gun, further freeing up the hands of the marksman. Pratt expounded the benefits of his invention in a 1916 patent:

> The gun is automatically aimed unconsciously and incidentally to the turning of the head of the marksman in the direction of the target. In self-protection, one immediately, instinctively turns the head in the direction of attack to see the enemy, or, in hunting, toward any sound made by nearby game. Thus the gun is automatically directed toward the mark in the course of the first instinctive movement. With the gun thus aimed, the only further operation necessary to fire the same is to blow through the tube and thereby expand the bulb and operate the trigger.[106]

Pratt's invention never reached the battlefield, perhaps as a result of the detrimental effect the firearm's recoil would likely have had on its wearer's neck. Nevertheless, it can be considered a forerunner to contemporary head- or eye-tracking technology that determines the direction of slaved guns or designates targets for guided weaponry by their operator's gaze. Yet a helmet can only be considered a visually coupled system when a control input is paired with a dynamic display. Via a continuous information loop, the head-mounted device issues commands to weapons, sensors, and computers that feed information back to it in the form of updated imagery (Figure 25). When scanning the world through sensors in this manner, operators experience a state of telepresence, peering out of the sensors as if they were doing so from their own body. Simultaneously, the system orients the operators' actions and observations through the superimposed display of automatic cues signaling

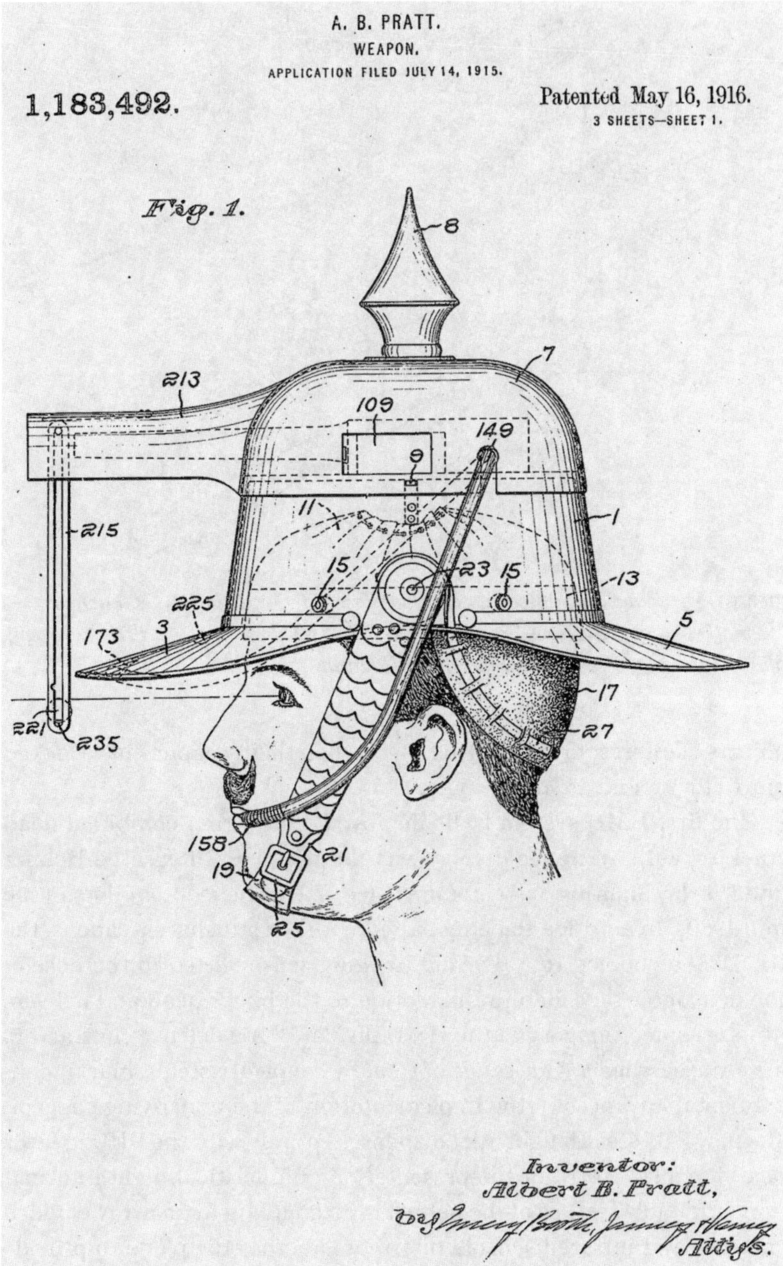

FIGURE 24. *Albert Pratt's helmet gun. U.S. Patent 1,183,492 (1916).*

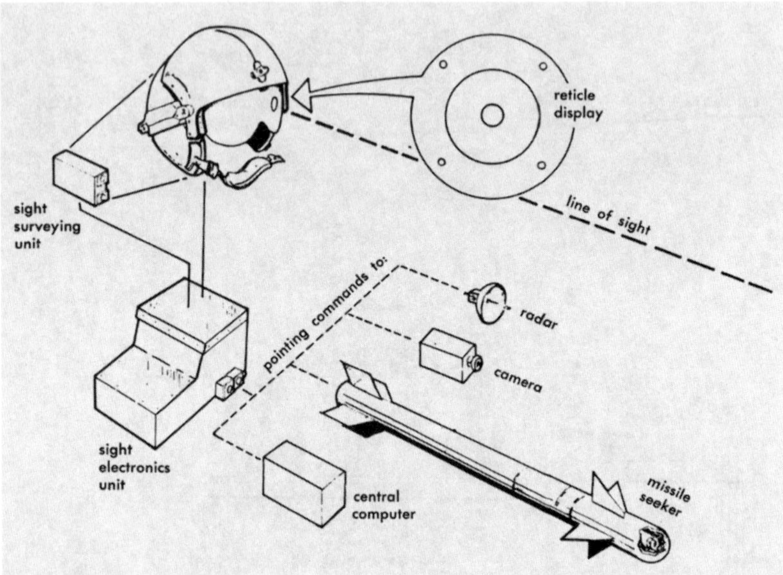

FIGURE 25. *Visually coupled system. Joseph A. Birt and Thomas A. Furness, "Visually Coupled Systems,"* Air University Review, *March–April 1974. Courtesy of Air University Press, Maxwell AFB, Alabama.*

objects of interest or instructing them when their weapons have locked onto a target and are ready to fire.

The first HMD system to be deployed widely that combined head tracking with an image display was the IHADSS (Integrated Helmet and Display Sighting System) for AH-64 Apache helicopter pilots in the mid-1980s. In addition to being paired with the gunship's weapons, the IHADSS is coupled to a thermal imaging sensor fitted to the nose of the helicopter that follows the motion of the pilot's head and delivers uninterrupted imagery to the HMD display.[107] Air defense contractors have since issued a wide range of visually coupled systems, marking incremental advances in the implementation of the underlying concept. The new F-35 stealth fighter jet dispenses entirely with the HUD in favor of a binocular HMD for the presentation of all critical flight information. A notable feature of the helmet is its coupling to an array of video cameras and infrared sensors distributed across the plane to provide a 360-degree view. The plane's computer seamlessly stitches together image feeds, presenting a continuous image to pilots as they move their heads, enabling them to look through the very frame of the aircraft in any direction.

Much like its parent aircraft, the F-35 helmet has had a notoriously troubled development, and it remains to be seen if it realizes its full promise in the operational environment.[108] Yet the helmet is emblematic of our singular position in the history of imaging, as characterized by the disappearance of the screen as a bounded surface for visual representation. Following Lev Manovich, we can think of a screen as any frame that exists within our natural, real space but whose surface opens up a vista onto another representational space. Even where perspectival representation achieved the optical artifice of a virtual space in continuity with real space, these continued to be phenomenally distinct, the fragile illusion of a unified space only holding so long as the viewer compliantly remained in a precise given position to the image. However, as imaging comes to simultaneously fill the entire visual field and synchronize itself with corporeal motion, "the real, physical space and the virtual, simulated space coincide," obliterating the prior distinction on which rested the conceptual screen.[109]

The automation of imaging necessary for the real-time visualization and exploration of three-dimensional virtual environments is a critical aspect of this development. Best known today for having spawned a large and profitable video games industry, these imaging techniques were originally devised through sustained military efforts to construct realistic aircraft simulators. Flight simulators appeared almost immediately after the advent of motorized flight with the construction of mechanical rigs that familiarized pilots with the controls and feel of aircraft. These grew in sophistication over the following decades, replicating cockpit instrumentation and simulating wind and weather conditions; by World War II, they had become widely adopted by combatant nations for pilot training. After the war, new technologies, such as the projection of video imagery from television cameras that roamed over model landscapes in response to the pilot's commands, further enhanced the realism of these simulators. Most decisively, computer image generation involving the dynamic display of three-dimensional physical models that characterize all contemporary flight simulators was introduced in the early 1970s.

The contribution of American computer scientist Ivan Sutherland's pioneering work to this last advance is fundamental. A doctoral student of the founder of information science, Claude Shannon, and a close colleague of computer vision pioneer Lawrence Roberts, Sutherland had already made his name in 1963 with the creation of an early graphical user interface that opened up an entirely new field of possibilities for human–computer interaction.[110] His next endeavor proved no less groundbreaking, with a "head-mounted three-dimensional display" that was likely the first virtual reality display system. Through a binocular display and a bulky overhead apparatus tracking the user's head motion that earned the device the foreboding moniker of "The Sword of Damocles," Sutherland was able to "present the user with a perspective image which changes as he moves."[111] Although the computer-generated graphics were just simple wire-frame geometric figures, the potential for creating synthetic environments was evident, and Sutherland promptly founded a company that went on to successfully commercialize military and civilian flight simulators.

Initially most computer simulators did not make use of HMDs, restricting themselves to using hemispherical domes or large cathode ray tube screens for the projection of three-dimensional images. However, a 1977 report from the Air Force's Aerospace Medical Research Laboratory outlined a scheme for a "Visually-Coupled Airborne Systems Simulator" (VCASS) that would use helmet-mounted displays for enhancing the realism and flexibility of digital simulation.[112] From this initial idea, the aspirations of the VCASS program grew rapidly, putting forward the most ambitious designs for a synthetic visual environment to date and far surpassing the objective of a realistic training simulator. In the final instance, the research team strove to create a combat-ready system that could simultaneously immerse users in a carefully optimized synthetic representation of reality and provide a virtual control interface for their actions.

From the outset, VCASS was intended to do more than merely provide heightened realism for pilot training. In visually representing the cockpit space as well as the navigational space beyond it, the simulator would allow for different instrumental and control configurations to be tested without the need for costly, time-consuming hardware alterations. Crucially, the program's researchers argued that VCASS should be installable in actual aircraft so that virtual representations of experimental control interfaces could be trialed against synthesized targets in real aerodynamic situations. In sum, VCASS was to provide "a self-contained airborne and ground-based man-in-the-loop visual simulator for the engineering of advanced weapon systems" that would combine "synthesized out-of-the-cockpit visual scenes and targets, a representative vehicle whose type can be altered, threat and weapon dynamics, flexibility of control and display configurations, and inputs from sensor or real world imagery."[113]

The VCASS program, which evolved over the course of the next few years, had an ultimate ambition of designing a virtual control interface that would definitively substitute itself for its physical equivalent in both training and fully operational environments. This expanded remit is reflected in the project's rebranding as "Super Cockpit" in 1986. No longer restricted to testing simulated interfaces that would eventually be physically instantiated, virtual reality was recast as the medium through which "the natural perceptual, cognitive and psychomotor capabilities of the operator" could be most effectively harnessed.[114] Superimposed over the real world with a one-to-one spatial registration (allowing for some transparency under daylight conditions), the three-dimensional display presented to the pilot would draw on "information from sensors, threat warning systems, terrain maps, and weapon delivery envelopes" to generate a streamlined representation of the operational environment, highlighting critical elements and features and enhancing the decision-making process (Figure 27).[115]

In addition to registering the position and orientation of the pilot's head through the HMD's motion sensors, the Super Cockpit was to track eye movement, monitor hand gestures, and respond to voice communications so as to allow control inputs to be issued from all these sources. The pilot would correspondingly receive not only visual but also tactile and auditory feedback issued by the "virtual world generator."[116] A further innovation proposed was the generation within the virtual cockpit of a so-called god's-eye display that would appear in the form of a miniature three-dimensional world representing the aircraft within its immediate surroundings. Interacting with this live cartographic interface with a gloved hand, the pilot would be able to initiate communications with friendly units as well as obtain information on displayed entities whose positions within the external view from the cockpit would be simultaneously highlighted.[117] The virtual cockpit, purportedly a more intuitive information display and flight control interface that any of its physical counterparts, could also be reconfigured at will in accordance

FIGURE 26. *VCASS helmet. National Archives and Records Administration (6414022).*

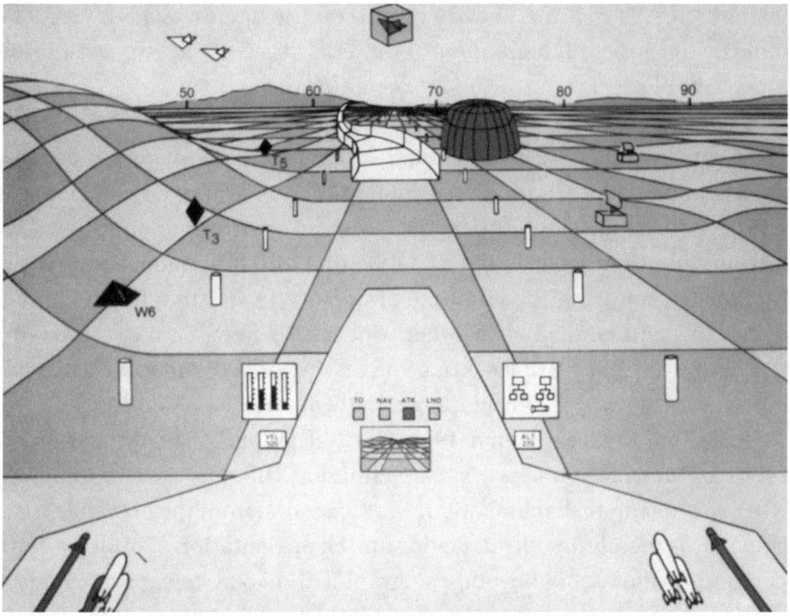

FIGURE 27. *Illustrative VCASS imagery. National Archives and Records Administration (6414024).*

with the mission, pilot preferences, or aircraft avionics. Finally, the pilot would not even necessarily need to "be present in the actual vehicle which he is piloting since with the appropriate data links a 'remote' virtual cockpit would provide the visual, auditory and tactile 'telepresence' cues as if he were located in the vehicle."[118]

The Super Cockpit program never did reach fruition; its funding was discontinued with the end of the Cold War. The full ambition of its virtual cockpit designs remains unrealized today in any field-deployed systems. Yet the legacy of many of its technological components can clearly be seen, notably in the synthetic vision systems that increasingly equip military and civilian aircraft. Such systems draw on geolocational positioning, sensor inputs, and digital terrain databases to generate real-time graphical simulations of the perspectival cockpit view from an aircraft in flight. Displayed via a stand-alone screen or integrated into HUDs or HMDs, synthetic vision can overlay graphics highlighting spatial elements of interest onto a video feed or a pilot's natural field of vision, or it can serve as a full navigational substitute in conditions of low visibility. UAVs that are heavily reliant on the narrow field of view offered by the optics of their onboard cameras also constitute particularly attractive platforms for the adoption of this technology.[119]

Above all, Super Cockpit constitutes a paradigmatic instantiation of a regime of machinic vision under which machines acquire independent faculties of perception and the human sensorium is rearticulated within new cognitive and perceptual architectures. Looping flows of information tightly bind man and machine into functional units, with the image occupying the vital role of the interface through which they command and control each other. Under this new regime, the traditional understanding of the image as resting on a distinction between representation and its object increasingly flounders.

In a short paper written in 1965, Ivan Sutherland, then at the head of DARPA's Information Processing Techniques Office, set out to imagine what the "ultimate display" might be. Reasoning from the principle that a display is merely an output device for the presentation of information, he conceived of going beyond the graphical display toward a "kinaesthetic display" that would stimulate all the human senses. This display would permit the translation of mathematical worlds encoded through a digital computer into full-body experiences. According to Sutherland, immersed subjects thereby might gain phenomenal access to such surreal objects as "negative mass," or "a triangle so built that whichever corner of it you look at becomes rounded." These fantastical constructs however give way to a more sobering concluding thought on the technology's logical end point: "The ultimate display would, of course, be a room within which the computer can control the existence of matter. A chair displayed in such a room would be good enough to sit in. Handcuffs displayed in such a room would be confining, and a bullet displayed in such a room would be fatal."[120] Under this arresting vision of the ultimate display, representation and its object coincide so fully under the imperium of information as to be wholly indistinguishable. Sutherland's final example revealingly recalls the elementary function of the martial image to fix objects for their targeting. Yet he goes one step further in encompassing the entire process within the single realm of imaging: the handcuffs immobilize and the bullet kills without ever leaving the confines of the display, such that in the conflicts of some distant future, being caught within the war machine's imaging capability may well signify being already at its absolute mercy.

CHAPTER 4

Mapping

> The elements of the art of war are first, measurement of space; second, estimation of quantities; third, calculations; fourth, comparisons; and fifth, chances of victory.
>
> —SUN TZU, *The Art of War*

Cartography has long served a role in the exercise of military power, where it has assisted with the strategic planning of campaigns, the design and placement of defensive fortifications, and the navigation of unfamiliar terrain. Yet until relatively recently, the use of maps in war was unsystematic and haphazard for all but the most specialized tasks and played virtually no part on the battlefield itself. Only at the turn of the nineteenth century did the military map emerge "as a central tool to handle the increased complexity and vast expansion of war," establishing itself as an indispensable medium for the information processing and transfer inherent to the command of modern armies.[1] A century later, the Great War set a new cartographic milestone with the comprehensive adoption of the scientifically surveyed map and its system of grid coordinates for the targeting of artillery and meticulous coordination of large-scale maneuvers. Finally, the Cold War's demand for increased accuracy and global reach in the projection of military force decisively brings the entire planet under a single architecture of geospatial address, recruiting in the process the full gamut of the earth sciences. Geospatially referenced navigation is concurrently automated, enabling weapon systems with synthetic powers of cartographic reasoning to autonomously locate and orient themselves in space. In the closing decades of the twentieth century, mapping is definitively assimilated into the informational paradigm of high-speed computation and dynamic data visualization, bringing geospatial representation into ever closer symbiosis with its users' cognitive operations.

A central feature of this historical narrative is the progressive striation of the globe, its geospatial extension enmeshed within a Cartesian grid of coordinates permitting any point on it to be located and related to all others. For David Harvey, this process of striation is inherent to the activity of cartography: "Cartography is about locating, identifying and bounding phenomena and thereby situating events, processes and things within a coherent spatial frame. It imposes spatial order on phenomena. In its contemporary manifestation, it depends heavily on a Cartesian logic in which *res extensa* are presumed to be quite separate from the realms of mind and thought and capable of full depiction within some set of coordinates."[2] In this sense, cartography fully participates in the standardization and gridding of space and time that, following Nigel Thrift, has made "different parts of the world locatable and transposable within a global architecture of address, so each and every part of the world [can] in theory be given an address" and the movement of objects within it "followed from location to location."[3]

Culminating in the advent of satellite geopositioning and computerized geographic information systems, this global architecture's conceptual foundations can be traced back to the oldest efforts at creating a rationalized system of cartographic representation. Already in the third century BCE, Eratosthenes had proposed that parallel and meridian lines could bring the positions of all locations in the known world under a single system of coordinates. A few centuries later, Alexandrian scholar Claudius Ptolemy adopted a similar principle of cartographic gridding and discussed at length the geometric principles of projection necessary to represent a spherical planet on a two-dimensional surface. The fifteenth-century rediscovery of Ptolemy's *Geographia* in Europe was a major spur to the development of a new scientific cartography that, together with the exploratory voyages of the age, would emulate and eventually surpass the achievements of the ancients in its reach and accuracy. In 1569, Gerardus Mercator devised a cylindrical map projection displaying the world's landmasses in shapes and proportions still recognizable to us in today's most familiar cartographic representations of the globe. Consciously looking back to Ptolemy, Mercator used "coordinate geometry as guarantee and means of a mathematically consistent plan and logically uniform set of rules" by which known geographic places could be located and future discoveries added.[4]

Because the Mercator map has frequently been evoked to challenge the claims of objectivity usually made by practitioners of scientific mapping, it is necessary to outline here the broad understanding of mapping underpinning the present chapter. In recent decades, critical geographers have insistently revealed the various aesthetic or ideological considerations that influence the production of maps, concluding that these can thereby no longer uphold the pretence of being neutral representations of reality.[5] The selections, omissions, classifications, and hierarchizations involved in the creation of any given map are in this view a reflection of the array of subjective orientations, representational biases, and rhetorical intentions of the cartographer and the wider sociocultural milieu. German historian Arno Peters famously attacked the Mercator world map for its use of a projection that distorts area sizes to the detriment of the third world, thereby revealing its inherent Eurocentric bias.[6]

Critical geographers are of course correct in asserting that one cannot remove mapping from the broader social and cultural setting in which it takes place, that in the final instance "maps are social constructions."[7] But the insistence on shifting our understanding of mapping away from the scientific to the discursive runs the risk of losing sight of the instrumental function of the map as a practical means for the "domination of space."[8] Indeed, the choices of geometric projection, scale, and features within a given cartographic representation are also guided by the specific practices of geospatialization in which that map participates. Mercator's original motivation for using the geometrical projection was to permit courses of constant bearing to be drawn as straight lines, a major innovation on the portolan charts then used by navigators.[9] The map can therefore also be conceived as a technology for the support and automation of cognitive processes in the manner of various other "material anchors" for conceptual representation or symbolic manipulation.[10] And while there is certainly no unalloyed functional determinism accounting for map design, such pragmatic considerations can bear heavily on it, all the more when cartographic representations are inserted into wider sociotechnical assemblages that crucially rely on them for their operation. Therefore, maps cannot be entirely reduced to their ideological or rhetorical effects—significant as they may be in certain instances—any more than they can be considered straightforward, neutral representations of reality. Rather, in many cases, they fulfill particular cognitive functions in specific operational assemblages and to which contingent but internally consistent criteria of efficacy will apply. In this sense, some maps will be better than others in facilitating the orientation, navigation, and decision-making of users within their environment.[11] Maps designed to be used for the strategic or tactical deployment and guidance of military force seem an evident fit for such an understanding. Of course, such maps undoubtedly direct and entrain their users toward certain apprehensions and actions in the world (for example, in framing a site as a target or a feature of terrain as cover from fire), but no more so than the other sociotechnical constituents of a given assemblage.

The Emperor's Map

While visual representations of geographic space have a long and rich history, maps' centrality in military practice is a comparatively recent phenomenon. David Buisseret tells us, for example, that the main English campaigns in France during the Hundred Years' War were conducted without any cartographic support and that local geographic knowledge was mainly acquired, as was typical at the time, through spies and scouts.[12] During the early modern era, maps began to play a significant role in armed forces' terrestrial and maritime navigation, the planning of sieges, and various aspects of military engineering. As such, cartographic representations were jealously guarded possessions of considerable strategic import. The French Marshal de Vielleville could thus declare in the sixteenth century that "a military commander must no more move without a map than a pilot or galley captain, unless he wants to court disaster."[13] However, most early modern maps' partial and unsystematic character considerably restricted their operational and tactical uses. As late as the eighteenth century, cartographic representations still notably lacked such features as contour lines denoting the elevation of terrain.[14]

The increasing military importance attached to maps is due in no small measure to the development of scientific cartography governed by strict geometric principles and standardized representational practices. As featured in previous chapters, the Renaissance rediscovery of triangulation provided a means of effectively measuring the distance to a remote visible point in space through the application of trigonometric principles. This surveying technique proved invaluable to the exercise of mapmaking because any measurement between two discrete points obtained in this manner could thereupon serve as the baseline for the triangulation of a further point. By painstakingly layering a mesh of contiguous optical triangles across an area, vast distances could be mapped with unprecedented accuracy. Dutch mathematician Willebrord Snellius is generally credited with the first systematic use of triangulation networks in 1615, but the most resounding demonstration of their value to cartographic ventures came with the famous Cassini map of France.

Under the stop–start impulses of the French monarchy and private enterprise, four generations of the Cassini family undertook an exhaustive triangulation of the national territory between 1683 and 1790, on the basis of which was produced a complete map of the country at a scale of 1:86,400. Far and away the most accurate and detailed cartographic representation of any country at the time, the Cassini map was only the fullest realization of a common ambition among early modern rulers to improve the geographic knowledge of their territories so as to more effectively govern them.[15] Not least among the concerns of those states was their preoccupation with the defense of their possessions from the predations of other states, on occasion entering into conflict with the designs of their civilian mapmakers. French generals would, for instance, vociferously argue that the commercially available Cassini map should not contain accurate representations of the country's borders and coasts that might provide hostile states with valuable military intelligence.[16] Yet as groundbreaking as the Cassini maps were, their small scale and approximations of relief did not provide the topographical detail necessary for the strategic movement of military forces, let alone preparation for battle.

The British Ordnance Survey, as its name readily reveals, grew out of more explicitly military imperatives. After early attempts to map the Scottish Highlands in the mid-eighteenth century as part of the efforts to suppress seditious clans in the wake of the Jacobite rebellion, plans for a national survey received a major boost from the fears of invasion prompted by revolutionary upheaval in neighboring France. Beginning with the vulnerable southern coast, the Board of Ordnance embarked on a triangulation of the British Isles that was completed by 1853.[17] The scale of the final map was chosen for its convenience in moving infantry, reflecting its intended military purpose.[18]

If the Ordnance Survey originated in the fears over Napoleon's designs over England, the French emperor would be directly responsible for the first systematic use of maps in the conduct of war. Frequently lacking support within military hierarchies and consequently susceptible to neglect and dispersion in peacetime, specialized geographical units had begun to assert their value to military commanders in the eighteenth century.[19] Napoleon's revolution of the military art, characterized by geographically expansive campaigns and innovative coordination of discrete armies in the field, established the indispensability of topographically accurate maps. The emperor attached great importance to such cartographic representations and devoted unprecedented resources to their production. The so-called Carte de L'Empereur, a uniform relief map of Europe at the 1:100,000 scale, the sole copy of which was eventually lost in the retreat from Russia, was his most celebrated commission. However, the countless local topographic surveys drawn up by military engineers were just as important to his armies as they maneuvered across Europe and accumulated spectacular successes on the battlefield (Figure 28).[20]

In addition to supporting the spatial orientation of officers in the field, the military map served the higher echelons of French command "as a media for time-space modelling and analysis of strategic movements."[21] Anders Engberg-Pedersen recounts that Napoleon's staff would inscribe the map with pins, colored pencils, and compasses to indicate the latest known positions and movements of friendly and enemy forces and to highlight important topographical features, as well as to calculate distances between discrete points. Staff members could virtually play out various strategic and tactical scenarios until a final decision was arrived at and orders were issued. As "a means of visualizing and managing the future," the Napoleonic map can be thought of as "the central part of an information-transformation system."[22] It is perhaps no coincidence then that Baron von Reiswitz's *Kriegsspiel,* the first modern war game, appears at this very time, with its terrain maps and complex set of rules designed to simulate the imperfect in-

FIGURE 28. *Napoleonic military geographers. Henri Marie Auguste Berthaut,* Les Ingénieurs Géographes Militaires, 1624–1831 *(Paris: Imprimerie du Service Géographique, 1902).*

formation and limited means of communication available to military commanders.[23]

After the conclusion of the Napoleonic wars, European and American militaries became engaged in increasingly voluminous cartographic production throughout the nineteenth century, notably in the new colonial territories that had to be charted and apportioned between rapacious empires.[24] The scale and duration of the military campaigns during the American Civil War provided yet another impetus for the acquisition of reliable maps. By 1864 the United States Coast Survey and the Army's Engineering Corps were delivering around forty-three thousand maps to the Union Army on an annual basis.[25] In 1908, Captain Henry Eames could proclaim that "hand in hand with the development of the science of war has advanced the science of Topography; and as war emerged from the domain of art into the cold, true atmosphere of science, soldiers have placed more and more reliance on the cartographer's representation of the theatre of operations."[26] In practice, topographical maps of sufficient detail for the planning of military operations were then still frequently produced in a hurried, ad hoc manner that restricted them to a rough survey of the main natural and artificial features of local terrain detached from any wider cartographic or geodetic referents.[27] Such approximations would become wholly inadequate with the outbreak of World War I, a war of static fronts and artillery that came to rely on an unprecedented degree of cartographic precision and that would firmly place military mapping on a scientific footing from which it has not since departed.

Map Shooting

Until the end of nineteenth century, an active battlefield was usually small enough to permit military commanders to view its totality with their own eyes. The term of art of coup d'oeil, as found in the writings of such illustrious figures as Frederick the Great, Napoleon, and Carl von Clausewitz, thus referred to the innate ability of the skilled general to survey terrain or a military situation at a glance and determine the relevant tactics to be used.[28] Commanding officers could communicate maneuvers to their subordinates by pointing at visible terrain features, and artillery guns were still principally aimed at targets that lay in their direct sights.[29] Maps were accordingly limited to strategic and operational purposes, advantageous in guiding the movement of troops throughout a campaign, identifying a location in which to do battle, or determining the best positions for forces to occupy on a given terrain but essentially devoid of tactical value once combat was initiated.

The rapid transformation of warfare would greatly expand maps' role in command during the twentieth century. From World War I onward, the dramatic extension in the scale of battle necessitated new cartographic visualizations to provide military commanders with a synoptic oversight. Simultaneously, maps had to be distributed down the chain of command for the implementation of tactical schemes. The needs of artillery, particularly for the exercise of indirect fire aimed at targets beyond the line of sight, set new standards of cartographic accuracy. Finally, the drastic alteration of terrain resulting from the conduct of military operations, in particular the cumulative effects of artillery, mechanized infantry, and aerial bombardment, required that maps be frequently updated to reflect the changes in the landscape and disposition of enemy forces within it. As Roy MacLeod puts it, warfare in the new century "placed new demands on the craft of military mapmaking, transforming the practice from one of relatively passive information gathering, to one of active and changing use. These changes led to the introduction of a new kind of 'information warfare,' combining traditional practices of engineering and artillery, with specialist skills in the use of observation and experiment."[30]

Both the available cartographic resources and prevailing geographic practices at the Great War's onset were wholly unsuited to the military needs imposed by the conflict. On the outbreak of hostilities on the Western front, the French army on whose territory most of the fighting took place relied on an antiquated 1:80,000 map of the country and a few local maps at the 1:20,000 scale that each possessed distinct prime meridians.[31] While such maps were adequate to the opening maneuvers of the war, they simply could not meet the exigencies of the subsequent static war of positions in which artillery and indirect fire dominated. Indeed, the concentration of military exchanges in relatively fixed geographic areas and the need for accurate artillery fire against targets that could not be directly sighted induced two marked effects on the exercise of cartography.

The first of these was the stimulation of surveying efforts of unprecedented intensity and the associated production of regularly updated maps for the combat areas.[32] These maps would all adopt a single common meridian and come in a wide variety of scales to suit the array of specific operational demands. It is estimated that over the course of the war, Britain printed thirty-four million war maps, France issued over thirty million, and Germany produced an astounding 775 million over the two fronts in which it was engaged.[33] Vital to the planning and conduct of operations, continuously revised maps became the repository of geospatial information derived from the methodical cross-checking of ground surveying, aerial observation, sound-ranging operations, front-line reports, reconnaissance missions, and prisoner interrogations.

The second cartographic effect of the war was the systematic implementation of a rectangular gridding of maps, which permitted the unique geospatial coordinates of any point to be immediately read off. Although lines of latitude and longitude could already be used to identify unique locations, rectangular coordinates, combined with altitudinal information, greatly facilitated gunnery calculations. Peter Chasseaud observes that it took several years to realize the arduous task of resolving "different sets of geographical data of differing accuracies into a homogeneous spheroidal trigonometrical system" that trained gunners could work with.[34] The establishment of this system would eventually enable previously inconceivable uses of artillery. The technique of so-called map shooting is among its most striking achievements. Also known as predicted fire, the practice involves aiming (or "laying") guns by reference to a previously surveyed map, allowing trained gunners to translate the coordinates of any point into firing instructions for its targeting. The diffusion of map shooting led to a number of novel tactical schemes being devised during the Great War, such as the creeping barrage, in which batteries of artillery coordinate their shelling according to a predetermined timetable of successive lines ahead of advancing infantry (Figure 29).[35] In this light, the claim by the head of survey efforts for the British Expeditionary Forces that "a map is a weapon" takes on its full import.[36]

For Chasseaud, the mapping effort during the conflict participated in the emergence of a new "battlefield geometry": "A new geographical information system had been created, providing a sophisticated three-dimensional fire-control database or matrix of the battlefield. . . . In effect, the battlefield had been digitised; every point could be rendered in metric or yard rectangular coordinates."[37] Admittedly, gridding was implemented across the front in a fragmentary manner, resulting in localized grid referents that could not be easily stitched together. Only the French military succeeded in attaining the goal of an encompassing

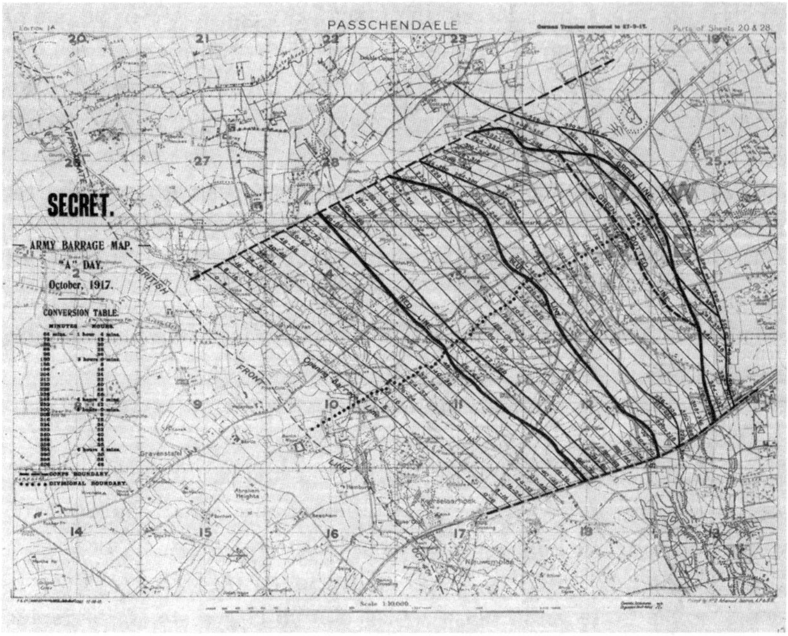

FIGURE 29. *A 1:10,000 scale creeping barrage map, Passchendaele, 1917.*

system that "unified the entire western front into a single mathematical space" before the completion of the war.[38] Yet World War I undoubtedly represents a major landmark in the constitution of an "architecture of address" for the delivery of lethal force to discrete points in space, the reach of which would only extend itself throughout the rest of the century.

Indeed, the intense cartographic gridding of the battlespace for the purposes of navigation and fire control resumed with World War II, incorporating a range of innovations in topography and aerial photogrammetry during the interwar period.[39] The unprecedented logistical and operational demands of a modern industrial conflict spanning entire continents fueled an insatiable demand for geographic information and its cartographic representations. Whereas World War I had been defined by a static front, the new conflict was characterized by movement, transporting troops into unfamiliar terrain. The advent of strategic air bombing on an industrial scale further demanded detailed and up-to-date knowledge of targeted areas lying deep into enemy territory.

The coordination and directing of artillery continued to be one of the primary functions of military cartography. Combined with the widespread adoption of portable radio, gridded maps now routinely enabled "the fire of a corps or divisional artillery to be immediately concentrated and swung about the battlefield as from one 'grand battery'" in a manner that had only been occasionally achieved with fixed-line telephony in World War I.[40] The U.S. military particularly sought a uniformity of grid zones so that their artillery could use the same mathematical tables and coordinate conversions everywhere. Confident of their superior calculational efficiency, American artillery could self-impose a delay of no more than thirty seconds between a command center's receipt of target information and the relaying of final coordinates to the field.[41]

In total, the combined efforts of the Germans, Russians, British, and Americans produced over three billion map sheets throughout the war.[42] Germany, accounting for almost half of this output, once again demonstrated the most voracious appetite for cartographic representations. Yet the United States also achieved an intense state of geographic mobilization as the country found itself operating across two major theaters, including the vast Pacific area of which it possessed limited prior knowledge. The U.S. government ended up employing 40 percent of the membership of the three national geography associations between 1942 and 1945, half of whom were engaged in cartographic work.[43] The Aeronautical Chart Service photographed a total of 40 million square kilometers—equivalent to a quarter of the earth's land surface—for the purposes of mapping.[44] By the end of the conflict, the newly formed Army Map Service had prepared forty thousand maps and issued around 500 million copies of them. Through the war, the United States military established a singular cartographic capability for mass production and global reach that was never subsequently demobilized.[45]

In the wake of the conflict, the United States was almost immediately locked into a tense showdown with the Soviet Union that would last for the next half century. This new confrontation entailed not only regional hot spots of conventional warfare scattered across the globe but also the possible use of uniquely destructive weapon systems across the breadth of the planet. A comprehensive and uniform geodetic system enmeshing the earth would therefore be realized to meet the prospect of a truly global conflict between nuclear powers, the culmination

FIGURE 30. *Poster, Map-Chart Division Headquarters, U.S. Army Air Services, 1942.*

of a vertiginous process of military escalation initiated at the beginning of the century—or as one American geographer pithily observed in 1954, "when war is conceived as total war, the geography of war is the geography of the earth."[46]

A Global Architecture of Address

In preparation for a potential clash with the Soviet Union involving long-range bombers and missiles, the United States launched into an intense geodetic and cartographic effort in the immediate aftermath of World War II. In 1947, the U.S. Army adopted the Universal Transverse Mercator system of map projection in combination with the Military Grid Reference System so as to plot the entire globe under a single grid of rectangular coordinates ranging from 100,000-meter grid squares to 1-meter grid squares.[47] However, the creation of a universal reference system resolved only one part of a larger problem because the reliability of any such grid is wholly dependent on its accurate layering on the geographical terrain being mapped—in this case the planetary body in its totality. Knowledge of the exact shape of the earth and the determination of the precise position of geographical points on it is therefore crucial, requiring the resources and expertise of the science of geodesy.

Geodesy (from the Greek word for "dividing the earth") has a longstanding history that can be traced back to at least the third century BCE. Greek mathematician Eratosthenes then produced remarkably accurate calculations of earth's circumference through the observation of the shadows cast by the sun in different locations. Notwithstanding its venerable forbearers, geodesy only became a sustained focus of study in the early modern era, galvanized not only by the new scientific method and its instruments but also by the practical necessities of Western exploration and colonialism. While the curvature of the earth can be ignored for most purposes when undertaking the mapping of localized areas, it becomes crucial for the accurate measurement and calculation of distances and positions over greater expanses. Because the true shape of the earth is not that of a perfect sphere but rather closer to an oblate spheroid (that is, flattened at the poles with a bulge around the equator), late eighteenth-century geographers began to make use of reference ellipsoids that approximate this shape for their geospatial measurements and representations. Combined with a set of reference points, such an ellipsoid generates the geodetic datum supporting a chosen geographic coordinate system.[48]

Because of the earth's uneven shape, geographers used different reference ellipsoids to generate more accurate geodetic systems across distinct regions, a practice that became problematic with the emergence of new demands for navigation. German rocket scientist Wernher von

Braun observed that the errant targeting of V-2 missiles at London had revealed a misalignment of several hundred meters between the national datums used for the mapping of Great Britain and France. In the eventuality of a nuclear war between the United States and the Soviet Union, the navigation of aircraft and missiles would require an accurate global datum to achieve the desired planetary reach of weapon systems. Such a task would be all the more taxing for the fact that the physical irregularity of the earth's shape is compounded by local variations in the globe's gravitational field. A satisfactory resolution of this problem eventually required the mathematical formulation of a highly uneven geoid in place of the smooth geometry of a reference ellipsoid. The determination of the "true physical figure" of the earth, which necessitated a vast array of measurements and observations, would only be attained through a "general mobilization of the earth sciences" under the aegis of a military-scientific complex intent on making global geopositioning a reality.[49] The U.S. Defense Mapping Agency's pedagogical *Geodesy for the Layman* (1984) underlines this historic convergence of scientific and military interests:

> In the past, military geodesy was largely involved with the practical aspect of the determination of exact positions of points on the earth's surface for mapping or artillery control purposes while the determination of the precise size and shape of the earth was a purely scientific role. However, modern requirements for distance and direction require both the practical and scientific applications of the science to provide the answers to problems in such fields as satellite tracking, global navigation and defensive missile operations.[50]

In 1964, the head of the Army Map Service, Robert C. Miller, made clear that his agency was now "primarily concerned with the various existing local datums and in reducing their number until all map information and geodetic data can be referred to a single unified world datum," at which point "the ultimate in a world system will have been reached."[51] The Department of Defense's first attempt at such a global geodetic system came in 1960 with the World Geodetic System 1960 (WGS 60).[52] With WGS 60 having demonstrated proof of concept, subsequent improved iterations followed in the form of WGS 66, WGS 72, and WGS 84. An updated version of the latter is still in use today, overlaid with the Military Grid Reference System adopted by all NATO militaries.[53]

The development of space-based satellite navigation systems has simultaneously enabled and benefited from geodetic advances. The Global Positioning System (originally Navstar GPS) is the most famous and ubiquitous of these systems.[54] Initiated and maintained by the Pentagon at a cost of over $35 billion to date, orbiting GPS satellites "lay a grid over the totality of the Earth's surface" and "mark every position on that grid with a real-time address, a unique and singular geotemporal code."[55] GPS now maintains levels of accuracy down to a few centimeters, depending on the receivers and techniques used, and is not restricted to laying a flat system of Cartesian reference coordinates on the surface of the planet in the manner of longitudinal and latitudinal measurements. It also enacts a full volumetric striation by locating receivers within a three-dimensional space of coordinates that extends from sea level to the upper reaches of the atmosphere.[56]

Although its global reach is unprecedented, GPS has some clear technological precursors. Most notably, radio navigation systems could already obtain a geopositional fix on a reference map by measuring precisely the difference in the time delay between the reception of synchronized signals from two or more radio stations. The British first introduced so-called hyperbolic navigation in 1942 to assist in the wartime navigation of their aircraft and ships. The system was dubbed Gee ("G") for "grid" because its objective was, according to the chief of British radar efforts, to "unfold [an] electronic grid over Germany."[57] The United States produced over seventy-five thousand receivers for its own LORAN (Long Range Navigation) system, which operated at the lower frequencies and longer ranges necessary for the Atlantic and Pacific theaters of World War II. Both systems involved reference maps crisscrossed with hyperbolic lines, representing discrete values in the time delay between radio signals, that wove together a veritable lattice of electronic coordinates (Figure 31). By reading off the values from signals provided by a radio receiver, navigators would find their present longitudinal and latitudinal coordinates at the intersection of two hyperbolas. LORAN and similar hyperbolic radio systems continued to play a major role in assisting the navigation of both military and civilian platforms throughout the Cold War until the advent of space-based satellite navigation systems.

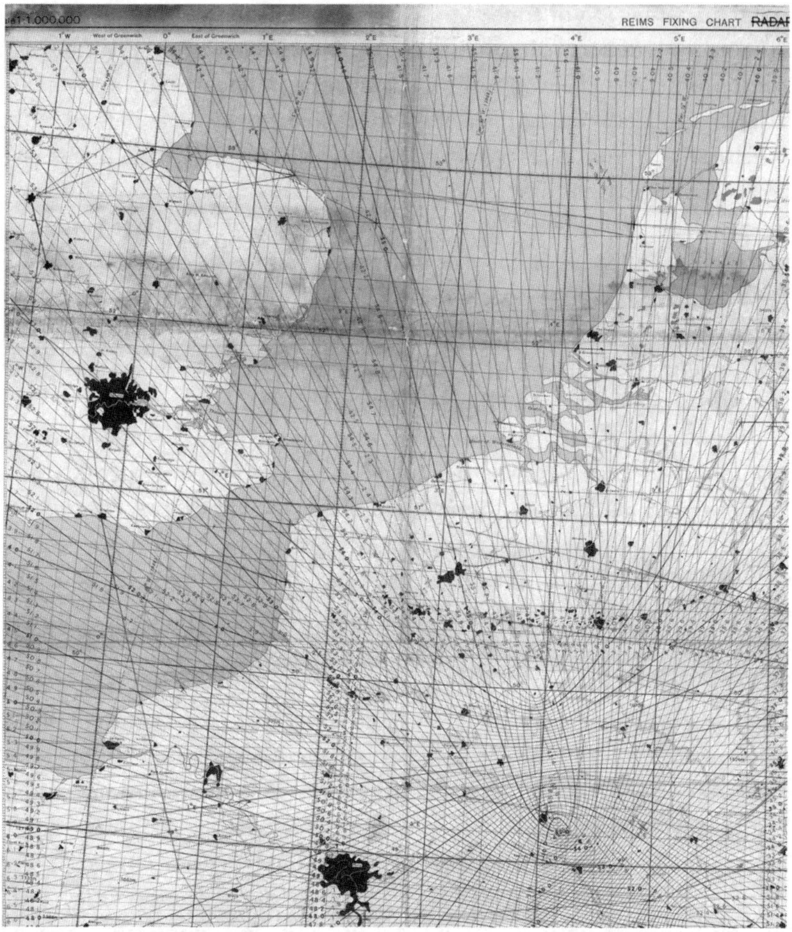

FIGURE 31. *Gee lattice chart, Reims chain, 1944. National Air and Space Museum, Smithsonian Institution.*

As illustrated by radio navigation and the previously discussed cases of sound ranging and radar (see chapter 2), the history of geopositioning is intimately linked to that of timekeeping, the gridding of space advancing in tandem with the increasingly fine striation of time. The eighteenth-century invention of the marine chronometer, a portable clock able to accurately keep time aboard a ship, thus solved the enduring navigational problem of determining longitudinal position.[58] By the early twentieth century, the synchronization of clocks through the radio transmission of time signals minimized the drift of mechanical timepieces and accordingly improved the precision of geopositioning. Yet it would require a literal quantum leap in chronometry to make satellite navigation systems a reality. All GPS satellites, equipped with synchronized atomic clocks set to the ultrastable oscillation in the energy states of nuclear particles, keep time within an accuracy of under a second in 300,000 years.[59] On the basis of the known fixed speed of radio waves (i.e., the speed of light), the delay necessary for a time-coded signal to reach a GPS receiver allows for the precise calculation of its distance to the emitting satellite.[60] By combining the distance measurement to at least four satellites in direct line of sight, the exact three-dimensional location of the receiver can be determined.

While the GPS program would not be declared fully operational until 1995, the 1991 Gulf War was the occasion for the full-scale demonstration of GPS's value as a military technology. Indeed, GPS is commonly held to have "played an important, if not crucial, role in the success of Operation Desert Storm."[61] Assuming featureless, trackless desert terrain to be an insuperable obstacle to navigation, the Iraqi military had concentrated its forces along the Kuwaiti–Saudi border, where preexisting roads and pipelines would provide the landmarks presumed necessary for any coalition advance. With the assistance of GPS, the coalition was instead able to move swiftly across the desert, barely impeded by sandstorm conditions, and attack the vulnerable western flank of the Iraqi army to score a spectacular victory.[62] In addition, commentators have commonly singled out "precision-bombing, artillery fire support, the precise positioning of maneuvering troop formations, and certain special forces operations . . . as either having been made entirely possible, or greatly enhanced by GPS."[63]

These impressive results were achieved despite the U.S. Army hurriedly acquiring just over four thousand GPS receivers in time for the 1991 Gulf War, almost 90 percent of which were commercially available devices. By contrast, at the outbreak of the 2003 Iraq war, the Army claimed to hold over 100,000 receivers.[64] If only a fraction of the 1991 guided weapons relied on GPS, the U.S. military has since implemented the navigational technology, at relatively low cost, to turn existing unguided munitions into smart bombs that home in on a geographic grid coordinate. Combining a GPS receiver with inertial guidance, Boeing's JDAM (Joint Direct Attack Munition) had shipped over 300,000 units

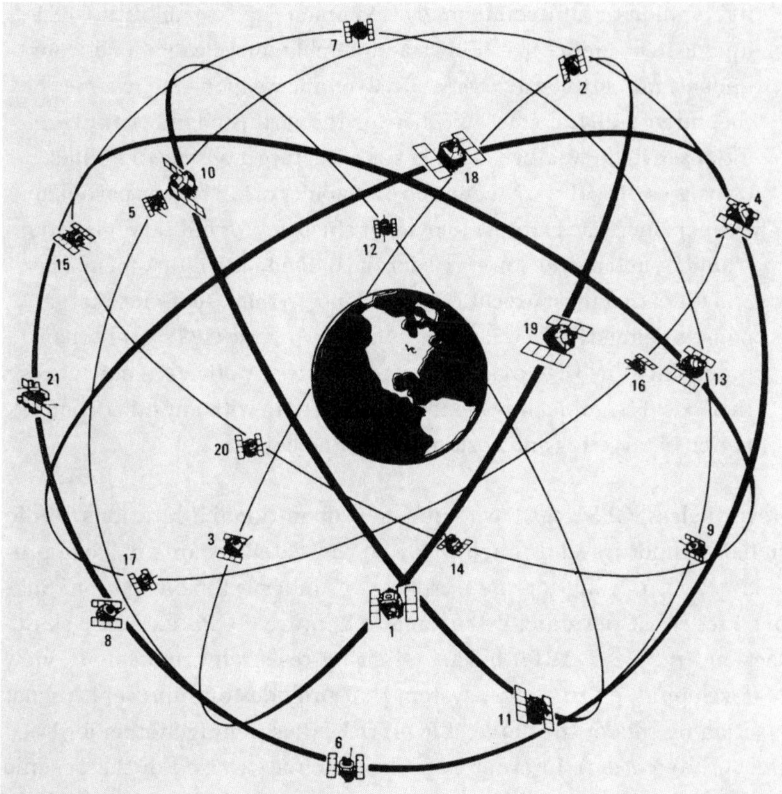

FIGURE 32. *GPS satellite constellation as envisioned in the mid-1980s. From R. L. Beard, J. Murray, and J. D. White, "GPS Clock Technology and the Navy PTTI Programs at the U.S. Naval Research Laboratory,"* Proceedings of the Eighteenth Annual Precise Time and Time Interval (PTTI) Applications and Planning Meeting *(1986). Image scan courtesy of William Rankin.*

by 2017 and played a prominent role in the postmillennial wars in Afghanistan, Iraq, Libya, and Syria.

Today GPS is a mission-critical component of military operations, vitally enabling the present American vision of high-precision and low-risk warfare. As a task force of the Defense Science Board expounded in its 2005 report on "The Future of the Global Positioning System":

> For military missions, GPS provides an unparalleled force-enhancement tool. GPS aids in all aspects of military combat operations because of

> its common-datum, common-grid, common-time capabilities. GPS is unique in its ability to establish an unambiguous correlation in four dimensions between a target and a dynamic weapon system aimed at that target—all the time, anywhere on the earth, and under any conditions of light, weather, or other source of target obscuration. This translates directly into increased probability of kill for any particular weapon, increased force employment efficiency for military mission planners, and overall lower risk for the individual military members and units that must execute the missions. To the extent that a target point is defined and a weapon is guided by precise GPS signals, the probability that the target will be hit despite any other circumstances that exist is significantly higher with GPS than with any other combination of targeting and positioning technologies.[65]

Nevertheless, GPS does have important operational limitations. Radio signals cannot travel through water, so GPS satellites are unable to provide geopositioning for the purposes of underwater navigation. Submarines must periodically surface to acquire a GPS fix and risk detection. In 2015, DARPA began soliciting research proposals in view of developing "an undersea system that provides omnipresent, robust positioning" under the name of POSYDON (Positioning System for Deep Ocean Navigation). In order to realize such a striation of the oceanic depths, the system would entail the distribution of "a small number of acoustic sources, analogous to GPS satellites, around an ocean basin" with positioning provided by "measuring the absolute range to multiple source signals."[66]

The drive to further extend volumetric striation to spaces inaccessible to satellite geopositioning signals includes the subterranean world. In 2010, DARPA budgeted $4 million for investment into a program called Transparent Earth. Through the creation of sensors and algorithms that can generate three-dimensional maps in real time, the program aspires to display "the physical, chemical, and dynamic properties of the earth down to 5 km depth, including natural or man-made structures at militarily-relevant spatial scales."[67] In a manner analogous to radar and sonar, Transparent Earth would use very low frequency radio or sound waves to penetrate the surface of the earth and interact with geological structures to produce measurable resonant signals. By harnessing the input from sensors attuned to these signals through a common data model, DARPA aspires to achieve an integration of "volumetric elements into a global three-dimensional picture of the earth's subsurface with variable spatial, temporal, and information resolution, allowing changes at local scales to propagate through both physical models and proximity rules to update the global picture"—in short, to produce a live map of the planet's crust in which any objects of military import can be located.[68] As Ryan Bishop points out, Transparent Earth and other such efforts to extend the martial gaze subterraneously are motivated by the proliferation of deep underground military bases, facilities, and tunnels, each a response to the increasing scrutiny and vulnerability of the surface of the earth.[69]

Orient and Destroy

Despite its prevalence, satellite navigation is only one available means for the delivery of lethal force to known geographic coordinates. Since World War II, militaries have developed a range of navigational and targeting technologies that make use of increasingly detailed geospatial information. Inertial navigation, commonly referred to as inertial guidance when it is used to direct a missile to an intended target, has its roots in the techniques of dead reckoning practiced by maritime navigators for centuries to calculate their displacement from a previously determined position through estimations of speed and course. Contemporary inertial navigation systems perform this task through an array of accelerometers and gyroscopes that register precise rates of motion and angular change. By rapidly computing the position and orientation of a platform from a known point of departure without the need for external reference, such systems can automatically issue any necessary instructions for the correction of deviations from a desired trajectory of travel toward a target destination.

In the closing years of World War II, the German V-2 became the first ballistic missile to make use of inertial guidance. The Allies swiftly appropriated the technology after the conflict's conclusion, and subsequent iterations provided the privileged means for directing intercontinental ballistic missiles (ICBMs) throughout the Cold War.[70] As a self-contained navigation system or black box, inertial guidance operates independently of weather conditions and does not necessitate the reception of any external signals prone to being intercepted, jammed, or impersonated. However, the accuracy of any such guidance system hinges on its inertial frame of reference, within which orientation and displacement are registered, being paired to the required geospatial data. During the Cold War, the flight paths of ICBMs had to be painstakingly programmed in advance, using their intended targets' known geospatial coordinates as well as the pertinent values of earth's rotation and gravitational field. As we have already seen, considerable efforts were expended to acquire both the necessary targeting intelligence and geodetic reference frames.

Although ballistic missiles were their first application, inertial navigation systems have since equipped a wide range of military and civilian platforms, including aircraft, ships, submarines, land vehicles, and even the ubiquitous consumer smartphone. In practice, however, inertial navigation systems suffer from spatial drift caused by the accumulation of errors in the gyroscopes' and accelerometers' measurements, resulting in an increasing loss of accuracy over time. While this drift is tolerable for high-yield nuclear ballistic missiles, for other applications external sources such as GPS are needed to make periodic recalibrations to the system. Consequently, inertial navigation is frequently combined with other techniques for geospatial orientation and direction to ensure the accuracy of its applications.

Terrain-referenced navigation, correlating the input from live terrain sensing with recorded topographic data to situate a given platform, is one such technique to have come to the fore. We find in a U.S. patent filed in 1946 the description of a "flight indicating instrument" whose basic operational principle prefigures all subsequent implementations of terrain-referenced navigation (Figure 33). The patent explains that a reconnaissance aircraft, following a predetermined course, altitude, and ground speed, was to photograph in advance the terrain to be flown over. Alternatively, the reference imagery could be "obtained artificially by preparing the picture strip of the terrain to be flown over, somewhat like a map."[71] The aircraft equipped with the navigational instrument could then run a projection of this film strip overlaid with live radar returns of the ground. By aligning the two images, the flight navigator would determine the aircraft's present ground speed, as well as obtain spatial bearings that might otherwise be unavailable in conditions of poor weather or visibility.

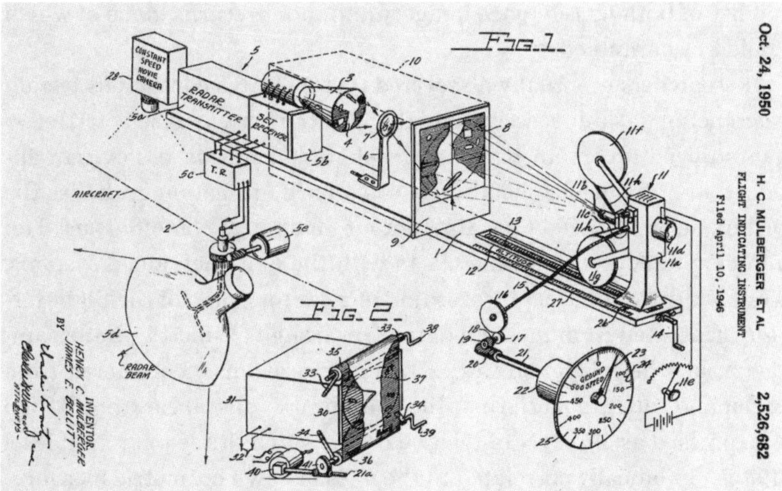

FIGURE 33. *"Flight indicating instrument." U.S. Patent 2,526,682 (1950).*

While the patent envisioned the proposed technology as a navigational aid for human operators, engineers promptly combined the same principles with cybernetic steering for the automation of missile guidance. By 1947, researchers were investigating the means for guiding long-range missiles via the referencing of onboard sensor input to known geographical features of the land to be overflown. Developed by Goodyear Aerospace for the U.S. Air Force, Automatic Terrain Recognition and Navigation (ATRAN) continuously compared observed radar returns to reference images stored within the guidance unit on a 35mm film strip.[72] In a process known as map matching, a video scan of the reference imagery would be algorithmically correlated on a real-time basis to a video signal generated by the radar sweeps, thereby determining any discrepancies necessitating automatic flight control corrections. The Army Map Service generated the reference imagery from handcrafted relief models that were illuminated and photographed from a vantage point simulating the missile's desired position relative to the real-world terrain. ATRAN was deployed on Mace cruise missiles in the late 1950s. One of its chief appeals was its autonomous geographic guidance, which reduced its vulnerability to jamming relative to alternative methods of radio control. However, ATRAN's performance was heavily dependent on favorable weather and light conditions, as well as the quality of both its reference imagery and radar returns, none of which could be guaranteed.

Researchers eventually discovered that ATRAN's continuous terrain referencing was unnecessary and that position checks at selected intervals sufficed if combined with inertial guidance. This less exigent application of the principle of terrain reference navigation underlies the technique of TERCOM (Terrain Contour Matching), commonly used on contemporary cruise missiles.[73] As with the aforementioned systems, TERCOM relies on the prerecording of a contour map of the flight terrain, generated from the data obtained through altimetry, photogrammetry, synthetic aperture radar, or lidar previously conducted from aerial and satellite platforms. The contour (or "digital elevation") map is expressed as a matrix of relative changes in altitude to ground level that is periodically correlated to the missile's own altimetric measurements as it proceeds along its flight course (Figure 34). Discrepancies between expected and actual measurements lead to the flight control mechanism making corresponding adjustments, compensating for any drifts in the inertial guidance. The increased accuracy in flight allows the cruise missile to maintain high speed at very low altitudes and even take a circuitous path to its target, reducing its exposure to radar detection and air defenses. Originally referred to as Fingerprint for its ambition to provide a unique identifier for terrain, just as dactyloscopy does for human digits, TERCOM was first conceived in the late 1950s but only matured sufficiently for field deployment toward the end of the 1970s.[74]

In the final instance, TERCOM's widespread use depends less on the guidance technology itself than on the extensive mapping infrastructure required to supply the necessary contour maps. According to one estimate in the early 1990s, the cost of generating digital mapping databases for TERCOM nearly matched the already considerable budget of the missile hardware.[75] More generally, increasing volumes of digital "earth data" have, from the 1980s onward, become essential for the operation of modern weapon systems and the accurate performance of the tasks of positioning, targeting, and navigation.[76] In the United States, the production of such databases falls principally under the aegis of the National Geospatial-Intelligence Agency (NGA), a successor organization to the Defense Mapping Agency (previously the Army Map Service), and various other historical government organizations tasked with the collection, exploitation, and dissemination of intelligence imagery.

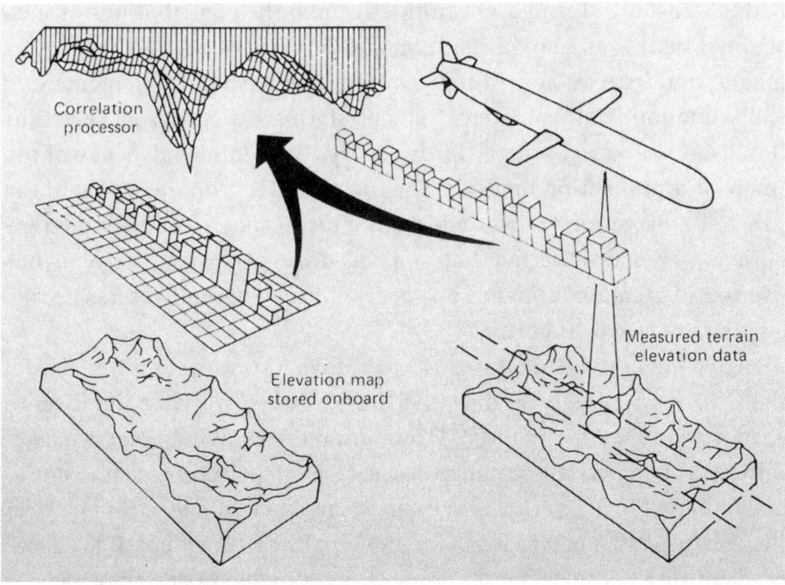

FIGURE 34. *Terrain Contour Matching. Joe P. Golden, "Terrain Contour Matching (TERCOM): A Cruise Missile Guidance Aid,"* Image Processing for Missile Guidance: Proceedings of the Society of Photo-Optical Instrumentation Engineers, *no. 238 (1980).*

Although classified, the NGA's annual budget is thought to be in excess of $5 billion and to have more than doubled since its latest renaming in 2003.[77]

The Digital Map Is the Territory

Underlying the past decades' advances in navigation and geopositioning is mapping's radical transmutation as it has been brought under the informational paradigm that arose from World War II. Over and above the digitization of its technological supports, the prior reconceptualization of the map as a communication channel and cognitive enabler has been decisive. With the emergence of the contemporary geographic information system has come a new mode of geospatial operability through which the exercise of military force is now increasingly being conducted.

In the wake of World War II, the cartographic establishment progressively adopted the new scientific conception of information that had taken shape during the conflict through the contributions of such seminal figures as Claude Shannon and Norbert Wiener.[78] The map, its maker, and its user accordingly could all be treated as elements of a single communication system. Influential American geographer Arthur Robinson is a key character in this story through his advocacy of the "map communication model."[79] Having played a prominent role in the American cartographic war efforts, Robinson spoke from personal experience when he asserted that "nothing during the past fifty years has exerted so great an influence on geographic cartography as has the occurrence of two world wars."[80]

After his appointment in 1942 to the head of the Map Division of the Office of Strategic Studies (the United States' wartime intelligence agency and precursor to the CIA), Robinson throughout his service committed to clear and efficient map design that purposefully eschewed propagandistic bias or artistic flourish. As an academic geographer after the war, he continued to insist on applying the highest possible degree of scientific rigor to cartography and sought to devise design principles and methods to that end.[81] Most significant, he conceived of the map as a medium of communication, whose main purpose had to be understood as the transmission of information to the mind of its user.[82] Such representations, he contended, would therefore have to be tailored to the scientifically determined features of human perceptual and cognitive faculties.

With the approach eventually known as the map communication model, Arthur Robinson and his followers were drawing chiefly on Claude Shannon's groundbreaking work on information theory, much of which was developed during his wartime mobilization.[83] One of the main progenitors of the digital revolution, the Bells Lab engineer is credited with the crucial conceptual breakthrough that any piece of information can be mathematically quantified and most efficiently stored and transmitted as a series of bits (binary digits), defined as the most elementary units of information.[84] One of the most important properties of Shannon's theory is that all communication systems, regardless of their specific technical characteristics, can be scientifically understood through the same abstract schematic (Figure 35). For any given message to be communicated from an information source (be it a human mind or a machine) to its intended destination, it must be converted into a signal for transmission across the specific medium of a designated

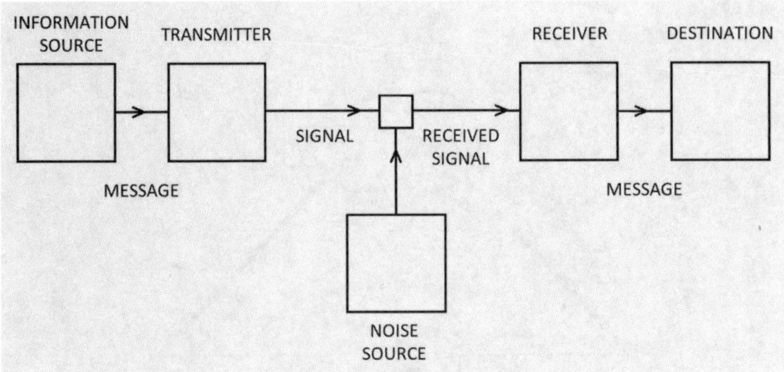

FIGURE 35. *Shannon's diagram of a general communications system. Adapted from Claude E. Shannon, "A Mathematical Theory of Communication,"* Bell System Technical Journal *27, no. 3 (1948).*

channel (with due attention paid to the persistent interference of noise degrading the signal and the corresponding need for redundancy).

For advocates of a scientific approach to cartography who saw the map as an information channel for the communication of geospatial data, Shannon's conception proved immensely attractive. According to geographer Joel Morrison, cartography was to be understood thereon as "the detailed scientific study of a communication channel."[85] The elaborate schematization of the map communication model proposed by Anton Kolácný in 1969 is clear evidence of information theory's influence (Figure 36). Unmistakably derived from Shannon's diagram, the model purports to represent the process of creation, transmission, and use of cartographic information within which the map serves as its medium. Along with an array of internal and situational variables that condition and orient the mental operations of cartographer and user, the model includes cybernetic feedback loops through which iterative cycles of the various processes bring the minds of the cartographer and user into convergence. As Kolácný put it himself, "The creation and communication of cartographic information is actually a very complex process of activities and operations with feedback circuits on various levels."[86]

Following from the communication model's conceptualization of the map as an informational channel between cartographer and map user (or "percipient," to use Robinson's terminology[87]), came the injunction

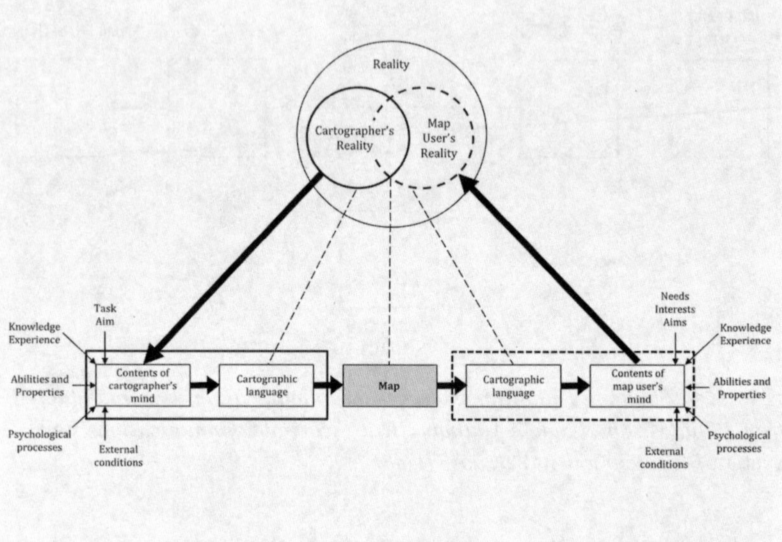

FIGURE 36. *Kolácný's model for the communication of cartographic information. Adapted from Anton Kolácný, "Cartographic Information: A Fundamental Concept and Term in Modern Cartography,"* Cartographic Journal *6, no. 1 (1969).*

for the development of a "cognitive cartography" that would apply "cognitive theories and methods to understanding maps and mapping and the application of maps to understanding cognition."[88] Robinson would thus claim in 1977 that "any thorough understanding of the field [of cartography] must involve a much greater penetration in which man's perceptual and cognitive processes are probed."[89] During the 1960s and 1970s, most of the research conducted into the map user's psychology was in the area of psychophysics, investigating visual stimuli and the perceptions they induce for the purpose of deriving general rules of map design. These studies were predominantly tied to a black box behaviorist model of the mind, even as it was being abandoned within the wider field of psychology.[90] Such an approach entailed a focus on stimulus–response mechanisms that promised to identify the "upper and lower bounds on the cartographer's freedom of design" but revealed little of the actual cognitive operations involved in map use.[91] Despite these considerable restrictions, the map communication model did nonetheless enact a convergence of cartography and the experimental study of communication and processing of information by the human mind.

In doing so, it brought the study and practice of mapmaking closer to Soviet geographer A. F. Aslanikashvili's conception of a "science which has as its subject cognition applied with the use of all the scientific-technical methods of cartography."[92]

The inherent limitations of the original communication model of mapping did nevertheless eventually become inescapable. Above all, it rested on the rigid assumption that the main purpose of a map was to transmit a predetermined message set by its cartographer to its recipient with the least possible distortion or corruption of the message. No allowance was made for the unforeseen applications or new meanings and interpretations that a user might bring to a map. Critical geographers highlighted this omission to impugn the positivist presumption of proponents of scientific cartography that maps are (or should be) disinterested neutral representations of reality with the essential task of cartographers being to increase accuracy and minimize error in these representations.[93] However, the emergence of GIS likely proved more significant in the decline of the original communication model. Within GIS, notions of the user as a passive recipient are simply no longer tenable—indeed, the very roles of producer and user tend to blur into one another. Yet the merging of cartography with the scientific study of human perception and cognition, as initiated by Robinson and his followers, not only endured but further deepened with the rise of GIS. Indeed, the latter shares with the map communication model a common adherence to the informational paradigm issued from World War II and its cyborg sciences.[94]

The outcome of an open-ended amalgamation of cartography, remote sensing, computer graphics, and database management, the geographic information system (frequently used interchangeably with "geographic information science") finds no single consensual definition. A diversity of applications and specific intellectual and institutional trajectories means that understandings and deployments of GIS vary considerably. For our purposes, we may nevertheless adopt the definition of "a system for capturing, storing, checking, manipulating, analyzing and displaying data which are spatially referenced to the Earth,"[95] each particular instance of which is a discrete manifestation of a wider phenomenon of geoinformatics. From the narrower cartographic perspective, the effects of GIS are nothing short of revolutionary, radically transforming the established understanding of the map and its uses.

Waldo Tobler is generally credited with laying the theoretical foundations of computer cartography with a 1959 article entitled "Automation and Cartography." Drawing on his time as a researcher at a defense software organization spun off from the RAND Corporation,[96] Tobler proposed that "the preparation of maps should be viewed as a complex data-processing system," considering the map in turn as the input, the data storage element, and the output of such a system.[97] Fifty years later, a contemporary account of the evolution of map seems to not only bear out but surpass this early vision:

> Over time, conceptual and technically driven developments in computer graphics, computation and user interfaces have begun to fundamentally transmute the role of the map from a finished product to a situation where the map is displayed within a visual toolbox to be used interactively for exploratory data analysis (typically with the interlinking of multiple representations such as statistical charts, three-dimensional plots, tables and so on). This changing conceptualization of the map is at the heart of the emerging field of geovisualization, which in the last decade or so has been one of the leading areas of applied cartographic research.[98]

Thus, for MacEachren and Kraak, maps are "no longer conceived of as simply graphic representations of geographic space, but as dynamic portals to interconnected, distributed, geospatial data resources" such that "the focus of both cartographic design and cartographic research now extends to problems in human–computer interaction and in enabling dynamic map and map object behaviors."[99]

Although largely neglected within both popular and academic accounts of its history, the development of GIS's key conceptual and technical components has been closely intertwined with military and security concerns. For Pickles, there is no doubt that "whether for weapons development, general surveillance or basic mapping exercises, GIS have been from the beginning a fully militarized activity, and in part this success stems from the abstract capacities and control systems generated by such data-handling and mapping capabilities."[100] Trevor Barnes underlines that Tobler's 1976 observation of a "gap between official governmental cartography and academic geographical cartography" gestures to the intense secrecy that surrounded advanced military-funded cartographic and geographic research such as his own during the Cold War, an opacity that to date declassification and the patient work of historians has only partially pierced.[101]

Some early military uses of GIS were certainly in evidence during the Vietnam War. In particular, the Hamlet Evaluation System (HES) was established to assist and assess the strategy of population "pacification" in the South Vietnamese countryside. Drawing on regular standardized field reports of questionable veracity, HES issued computer-generated maps purporting to capture Vietcong activity and changes in the local allegiances of some eleven thousand hamlets.[102] Yet the 1991 Gulf War plausibly constitutes "the first full-scale GIS war" in which space-based remote sensing generated detailed current maps of the theater and the geospatial technologies of GPS and cruise missile guidance featured prominently.[103] In addition to printing over 100 million sheet maps (three times the volume issued by either Britain or France during the entire First World War), the Defense Mapping Agency also produced six hundred all-digital map products for the conflict.[104]

The military adoption of GIS technology has since continued apace, becoming ever more closely interwoven with the conduct of operations alongside the increasing use of handheld computing devices and availability of data connectivity. The Command Post of the Future (CPOF) is one prominent program, designating a U.S. Army command and control software and hardware suite that integrates key GIS functionalities and serves as "a planning and mapping tool intended for collaboration between multiple echelons in a tactical environment."[105] CPOF began life in 1997 as a DARPA research project that drew on both military experience and expertise in the fields of cognitive psychology, human–computer interfaces, and information technology.[106] This convergence of fields is reflective of the increasing emphasis placed "on identifying and matching the cognitive needs of command-cell operators" in the development of battle command systems.[107] With the stated end goal of enhancing "situation awareness," understood as "the perception of the elements in the environment within a volume of time and space, the comprehension of their meaning and the projection of their status in the near future," such systems become tasked with actively supporting their users' cognitive processes.[108] CPOF is only one of many military ventures pursuing this objective today, of course. But with over twenty thousand terminals fielded by 2014 and noteworthy deployments in the

Iraqi and Afghan campaigns, it remains an especially instructive one, given the scale of its adoption and the central role that exercises in geospatialization occupy within it.

CPOF users primarily interact with the system via navigable digital maps available in two or three dimensions. On these maps, individual icons indicate the position of friendly and enemy forces as well as key events in the battlespace. Automatic updates through live data feeds from other geospatial information systems such as GPS-enabled Blue Force Tracking are supplemented by manual user inputs. In addition to strictly topographic representations of the area of operations, various digital map overlays can be summoned, providing information ranging from past incidents to the layout of infrastructure or the ethnic composition of the surrounding civilian population. Through the distribution of information across its terminals, CPOF is intended to provide a "common operational picture" among users that permits a "deep collaboration . . . at the thought process level that literally allows commanders, subordinates, and key battle staff to see what the commander is thinking."[109]

However, any notion that the system provides a definitive synoptic and hierarchically imposed oversight of the battlespace would be mistaken. In practice, as Caroline Croser has shown in her detailed study of the First Cavalry Division's deployment around Baghdad, CPOF actually induces a proliferation of maps that constitute "a nuanced and complex series of different versions of the battlespace" in response to the intrinsic multiplicity and mutability of the operational environment.[110] The battlespace thus "never resolve[s] into a single, definite picture," appearing instead as "constantly updated, fluid and always in the process of construction."[111] Rather than overwhelming users, this cartographic fragmentation and plasticity supports the cognitive navigation and processing of the vast information flows inherent to the conduct of late modern warfare. Alongside the global striations that enmesh the entire planet within a unified isotropic system of geospatial coordinates allowing any known position to be related to another, we thus also encounter localized exercises in mapping that generate fleeting geospatial representations through which causal inferences are made and attendant courses of military action are determined.

Although CPOF is merely one particular manifestation, it is nevertheless indicative of the profound ontological shift that mapping has undergone under the present informational regime. The map in its traditional understanding as a static, or at least punctuated, visual representation of space is giving way to systems for the dynamic interfacing of minds with repositories of geospatial data that exist in a state of continuous flux through their various data feeds and calculative operations. Subsumed under a generalized paradigm of computation and information processing that emerged in the middle of the last century, the map has come to be explicitly tasked with simultaneously being a visual representational form, a live communication channel, and a cognitive enabler, with the performance of each function to be methodically tailored to human capabilities and limitations. Mapping is thus coming to involve as much the charting of neural pathways and cognitive bottlenecks as the plotting of river tributaries and mountain defiles.

Nigel Thrift writes of "a world just coming into existence, one which is based on continuous calculation at each and every point along each and every line of movement," bringing with it new forms of spatial awareness.[112] It is in this light that we might best apprehend the recent efforts being expended on the acquisition of a capability for "mapping on demand," which makes use of roaming suites of sensors and real-time photogrammetry to automatically produce near-instantaneous three-dimensional reconstructions of an environment being navigated.[113] A related task exercising the field of robotics today is the performance of "simultaneous localization and mapping" enabling a computational agent to construct a map of unknown surroundings while simultaneously using that same map to locate its position within them. All of this suggests that geospatialization is increasingly becoming synchronous with the very exploration of an environment, hinting at a near future in which the creation of maps no longer merely precedes the conduct of military operations but has become a process entirely concurrent with them.

In his fragmentary short story titled "On Exactitude in Science" (1946), Jorge Luis Borges famously imagined a 1:1 scale map that was overlaid point for point on the imperial territory it represented. The production of an "art of cartography" held to have attained a state of "perfection," the very fidelity of this new emperor's map simultaneously rendered it devoid of any practical use.[114] Borges may well have diagnosed acutely enough the cartographic impulse from which such a map might issue, but he was arguably mistaken regarding its worthlessness. Or perhaps, more fundamentally, he was then still only able to conceive of such a cartographic endeavor as a single sheet of paper draped on the land it purports to definitively stand for. He did not—almost certainly could not—imagine the flickering layers of geospatial data and calculative operations ceaselessly binding the world into a meticulous correspondence with its dynamic digital models.[115] For this is indeed the shimmering landscape we now inhabit, its silicon valleys and algorithmic peaks already bounding tomorrow's battlefields.

CHAPTER 5

Hiding

Camouflage implies a seeing eye from which to hide.

—ADOLF PORTMANN, *Animal Camouflage* (1959)

The martial gaze's fateful ascendancy, through the development of its functions of sensorial perception, representational imaging, and geospatial mapping, has rendered increasingly vulnerable any entity caught in its sights. As weapon systems have ranged farther and faster, the relative safety afforded by distance has eroded. The protection offered by armor or fortification has been all but outstripped by the ratcheting up of destructive power. With the advent of modern sensors, digital information processing, and precision-guided munitions, military analysts began referring to the last century's closing decades as an "age of fatal visibility" in which observable targets face quasi-certain annihilation.[1] For Martin Libicki, visibility's equivalence with death can only mean a shift in the nature of warfare "from force-on-force to hide-and-go-seek."[2]

This trend was already discernible during World War I, with the constitution of the modern practice of camouflage responding to the unprecedented battlefield surveillance. Reflecting on the conflict in the interwar years, Ernst Jünger noted that "the endeavor to make oneself invisible grows" ineluctably as the perceptual reach of the war machine extends itself.[3] Yet even if for Gilles Deleuze and Félix Guattari "to hide, to camouflage oneself, is a warrior function" per se, traditional military cultures have long disparaged concealment, disguise, and subterfuge as marks of dishonorable cowardice and deceit.[4] Faced with the stark consequences of unprecedented exposure, the Great War's belligerents would have to cast off such engrained attitudes swiftly. Battlefield artifice would soon be understood, in the words of one British World War II camoufleur, as a legitimate form of "visual warfare" in which the field of perception is the very terrain of contestation.[5]

While not a primary function of the martial gaze, hiding forms its constitutive obverse by simultaneously counteracting its operation and providing a key impetus for its further elaboration. In adapting to the specific milieu it seeks to render itself indistinguishable from, camouflage must necessarily be attuned to the perception of the potential observer. Camouflage is, first and foremost, an anticipation of perception that strives to foil attempts at detection, identification, and pursuit. As the "art of visual deception," its efficacy therefore rests on "an understanding of the fundamentals of visual perception."[6] Nor can such an apprehension be limited to the senses of the human observer, vital as they may be; rather, it must extend to the workings of all the prosthetic organs of perception that survey the battlefield. The injunction, found in a 1943 psychology manual intended for the American armed forces, to "take advantage of another man's brain, use its own rules to deceive it, to make it perceive something that is not real" therefore holds more broadly for the array of sociotechnical assemblages of military perception that need to be confounded.[7] By necessity, camouflage draws on the same knowledge and techniques as the martial gaze, but it mobilizes them against it by exploiting their identified biases, lacunae, and blind spots. Military perception, in turn, is forced to address those weaknesses and further tighten its sensorial nets so as to recover the discriminant ability denied to it by concerted efforts in concealment.

This chapter begins by considering the origins of modern camouflage during World War I. In particular, it examines the crucial role of both the dramatic increase in the range and accuracy of gunnery and the advent of aerial photoreconnaissance in prompting the shift from conspicuous military displays to systematic practices of concealment and obscuration. Initially advanced by artists but increasingly supplemented, and eventually supplanted, by the contributions of scientists, an array of approaches were devised and deployed during the conflict, from blending and disrupting to screening and dazzling. In the following decades, most of these techniques would remain essentially unchanged, subject only to incremental adjustments in their design and implementation. Since the 1970s, the digitization of camouflage and algorithmic production of its patterns, the emergence of stealth technologies in response to the extension of sensorial perception across the electromagnetic spectrum, and sustained efforts to develop active forms of camouflage that can dynamically adapt to their immediate environment have fundamentally reinvented the arsenal of visual warfare. In addition to all the above efforts toward concealment, a complementary field of conduct seeks to draw, misdirect, and blind the martial gaze through decoying, simulacra, and bedazzlement. Finally, a set of more radical strategic responses to the ever-expanding means of military surveillance are coming to the fore today with the adoption of practices of generalized concealment, dispersal, and expendability that dissolve the linear and bounded space of the battlefield.

To Foil a Martial Gaze

Stratagems for concealment and avoidance of unwanted detection have an ancient history, the roots of which lie in the primeval activity of the hunt. In continuity with the various means of dissimulation and misdirection found among flora and fauna in the natural world, human predators have long sought to gain an advantage by disguising their own presence to their prey. Hunters thus learned to mask their sounds and smells, or to blend their visual appearance into their environment. In the conflicts between small groups of hunter-gatherers, such schemes could presumably also grant their practitioners a potentially decisive element of surprise over their enemies. Yet with the exception of scouting units tasked with discreet reconnaissance missions, systematic efforts at concealment by larger organized armed forces have, until recent times, been sporadic at best. Sizable armies on the march cannot be dissimulated easily, and actual battlefield engagements have generally involved dense concentrations of human bodies clashing at close quarters, offering little opportunity for concealment. On the contrary, many past military forces instead chose to loudly and visibly advertise their presence through drum cadences, marching songs, battle cries, standards, and distinctive attire.

Within the military institutions of the emerging European nation-states, fighting men's uniforms became standardized starting in the late seventeenth century, typically with vivid colors and ostentatious headgear that made their wearers highly conspicuous. The motives for these sartorial choices included bolstering the cohesion and pride of soldiers, appearing larger and more fearsome to their adversaries, and facilitating the recognition of one's troops in increasingly smoke-filled battlefields dominated by artillery and firearms.[8] Most certainly they were not intended to dissimulate the presence of troops to the enemy, with whom lethal contact continued to predominantly take place at short ranges until the closing decades of the nineteenth century. Thereafter, changes in the conduct of warfare, notably the rapid technological development of gunnery, began to challenge established vestimentary conventions. Barrel rifling and new propellant explosives dramatically increased the distances over which accurate fire could be directed, turning visibility into a dangerous liability. Furthermore, the advent of smokeless charges meant firing no longer immediately revealed the shooter's location, thus making the tactic of sniping potentially highly effective, as demonstrated by the irregular Boer forces at the turn of the twentieth century. In commenting on this feature of the conflict in South Africa, military strategist J. F. C. Fuller observed that "due to smokeless powder, the old terror of a visible foe had given way to the paralysing sensation of advancing on an invisible one."[9]

The Boer War also marks the first large-scale deployment of the British khaki-colored (or "drab") uniform, in direct response to guerrilla tactics. Khaki ("dust-colored" in Urdu) was originally introduced by a bespoke regiment of the British Indian army in 1848 and more widely adopted during the Indian rebellion of 1857. As an 1848 letter by one of the officers responsible for its original commissioning states, "The colour selected for their uniform was 'drab,' as most likely to make them invisible in a land of dust."[10] At the conclusion of the Boer War in 1902, the British army immediately designated khaki, albeit in a darker and greener hue better suited to European terrain, as the official service dress for all its forces, replacing the traditional regimental red coat. The American army followed suit the same year, and the German military switched from Prussian blue to field gray *(feldgrau)* in 1910. Other European armed forces, such as the French, were still hanging on to their traditional bright colors at the outbreak of World War I, but a jarring confrontation with the new lethal ranges of enemy fire soon relegated dashing uniforms to purely ceremonial settings everywhere.

Yet it was arguably the introduction of the airplane, with its singular ability to peer deep into the rear of opposing lines, that did most to drive the emergence of systematic practices of concealment and dissimulation during World War I. As we saw in chapter 3, armies assembled sophisticated logistical chains for the surveying of enemy positions from the air and the affixing of geographical coordinates to be targeted by devastating artillery fire. For Ernst Jünger, this quite simply meant that "a military position could no longer be held once detected by aerial reconnaissance."[11] It soon became conventional military wisdom that from now on, "concealment comes first, and protection is secondary."[12] This new injunction would eventually extend to warring nations in their totality as airborne targeting progressively widened to distant cities and industrial centers, giving rise to civil camouflage in World War II.[13]

From the outset, camouflage sought to evade detection from the air, and particularly the glassy stare of the camera, whose photographic images would be pored over by analysts seeking to extract valuable intelligence from them. According to one American camoufleur in the Great War, "At least 95 percent of our work had to do with perpendicular camouflage, the combating of photography on the part of the aerial camera."[14] The practitioner of camouflage therefore had to adopt an aerial perspective to determine how different objects appeared when viewed from above rather than on the ground. Camouflage designs that might seem glaringly conspicuous at close range could prove highly deceptive at a remove, and vice versa. Accounting for the changing light of the day, especially in casting telltale shadows that photographic interpreters would eagerly latch on to, was also crucial. Aerial photographic reconnaissance would commonly inspect an army's own positions and report back on any vulnerabilities or inadequacies in their efforts at concealment. Camouflage trainees were instructed in aerial consciousness through the application of their newly acquired skills onto model landscapes and an assessment of their effectiveness by photographing the results from a perpendicular perspective (Figure 37).

To further foil the photographic gaze, camoufleurs were acutely attentive to the medium of imaging's technical specificities and representational biases. For instance, aerial surveillance photography in the first half of the twentieth century almost exclusively relied on black-and-white emulsions. These images were much more sensitive to variations in tone and texture than to those in color, making them advantageous for photographic interpretation but also having important implications for the practices of concealment. Having noted this feature of aerial photography, the 1941 *Home Guard Camouflage Manual* advises

FIGURE 37. *Camouflage class in New York University, 1943. A student corrects the lapses in the camouflage of a model defense plant revealed by an overhead photograph. Photographer: Marjory Collins. FSA-OWI Collection, Prints & Photographs Division, Library of Congress (LC-DIG-fsac-1a34583).*

that, in selecting colors for camouflage designs, greens mixed with blue should be avoided in favor of yellowy greens mixed with brown, the latter combination producing tones closer to those found in nature.[15] The technical attributes of aerial photographs and the methods of their analysts continuously coevolved, not least in response to the increasing sophistication of camouflage efforts. Camoufleurs in the two world wars had to contend with the ever-growing arsenal of stereoscopic cameras, color filters, and various orthochromatic, panchromatic, and infrared films—each with its own particular sensitivities to different wavelengths of light—that became available for aerial reconnaissance and required corresponding adjustments in the practices of concealment. In Hanna Rose Shell's words, camouflage fundamentally obeyed "an adaptive logic of escape from photographic representation."[16]

The techniques of aerial concealment were not restricted to deceiving the photographic interpreter's eye, however, because they also had to contend with the viewfinder of the airborne bomber. Some important differences distinguished each gaze. For one, the angles of vision to the ground usually diverged—from the vertical perspective typical of the aerial photograph to the oblique one of the approaching bomber aircraft—and each made distinct demands on the dissimulation of vulnerable positions and sites. Each gaze also occupied discrete temporalities: the interpreter meticulously scrutinized photographs to uncover objects of interest before passing on the targets thus identified to the bomber, who in turn might only have precious seconds to recognize them from the air and release the munitions. Camouflage that failed to mislead the photographic analyst could still be effective if it successfully degraded the accuracy of the bombing, if only by sufficiently delaying its visual acquisition.[17]

Becoming Imperceptible

The early developments of modern military camouflage were the combined enterprise of artists and scientists. Initially the former much predominated, as painters, sculptors, and theater prop designers already possessed the requisite skills of artifice and illusion. As Abraham Rattner, a camoufleur in the U.S. Army during World War I, astutely put it:

> The camoufleur reverses the principle used by the artist who paints a picture representing an object in nature. The artist follows the principle of making a three-dimensional object appear as a three-dimensional object and he builds this appearance on the flat surface of the canvas. The camoufleur uses the plastic means of the artist in an opposite way because he is working on a three-dimensional form already existing—a cannon, a soldier, a pile of ammunition, an airplane, etc. and his object is to make these three-dimensional objects appear flat. The camoufleur must destroy, dissemble, make disappear a form.[18]

Zoologists and naturalists provided another obvious source of inspiration through their study of animal camouflage and the naturally evolved mechanisms by which predators and their prey conceal themselves from each other. But here again, the artist, rather than the dedicated scientist, realized the most significant breakthroughs at the outset. Present-day scientific studies of animal camouflage, histories of military subterfuge, and the literature of its contemporary practitioners all credit the foundation of camouflage's modern conceptions and practices to American painter Abbott Thayer. Beginning in the last decade of the nineteenth century, Thayer theorized some of the fundamental principles of animal concealment, thereafter vocally advocating their systematic application in the military sphere.[19] His most important and enduring contribution lies in the identification of two main modalities of camouflage: blending, in which a figure so resembles its background that it can no longer be distinguished from it, and disruption, whereby a figure is broken up through "the employment of strong arbitrary patterns of color which tend to conceal the wearer by destroying his apparent continuity of surface."[20] While both approaches serve the same purpose of concealment, they adopt seemingly antithetical means to do so. As Roy Behrens points out, "Blending is largely dependent upon a relationship of extreme similarity (between the figure and its ground), while disruption is based on the opposite tack, that of an excess of contrast (within the bounds of the figure itself)."[21] In broad schematic terms, blending thwarts the detection of an object within a given environment, whereas disruption interferes with the recognition of that object.

Blending strategies include matching a background's color (the self-evident rationale for the espousal of muted monochrome military uniforms in the first decades of the twentieth century) and countershading, where the underside of the camouflaged entity is lighter in color than its upper side so as to counteract the visual contrast caused by natural shadows. Yet as well matched to its background as a pattern may be (and of course mobile entities are likely to find themselves against a variety of backgrounds), camouflage may still struggle to conceal the familiar outline of a given object. However, bold markings and contrasts in color, luminance, or texture tend, in Thayer's own words, "to obliterate,—to cancel by their separate and conflicting pattern, the visibility of the details and boundaries of form."[22] The widespread embrace of disruptive patterns in the world wars, adorning countless artillery pieces, vehicles, and buildings, seems to validate their effectiveness. Although essentially restricted to scouts, snipers, and elite troops until the 1970s, such motifs have today become synonymous with military

FIGURE 38. *Royal Air Force light bomber with disruptive camouflage pattern, World War II. Copyright Imperial War Museums (CH 5554).*

uniforms through their myriad mottled compositions of spots, streaks, and patches in various shades of green, brown, and black. While operating according to apparently inverse principles, blending and disruption are actually generally complementary, and in practice, camouflage designs usually strive for some combination of both.

While the French were late adopters of subdued uniform colors, they proved to be pioneering in the wider techniques of camouflage that arose during World War I, thereby coining the term of art for posterity. Led by the painter Lucien-Victor Guirand de Scévola, the first French military section dedicated to camouflage was established in 1915. In addition to making use of the various talents of theater set designers and sculptors, Guirand de Scévola notably drew on the avant-garde movement of cubism that had come to prominence just before the war. The cubists famously rejected representation from a single perspective and light source and had a penchant for blending objects into their backgrounds. Recalling his earliest experimentations in camouflage, Guirand de Scévola would later write that "in order to deform totally the object, I had employed the means the cubists use to represent it," motivating the subsequent recruitment of "painters capable, by their very special vision, of changing the nature of any form."[23] Having established the value of systematic visual concealment and disguise, the French camoufleur ranks would rapidly grow, reaching about three thousand dedicated soldiers by 1918, alongside large civilian contingents laboring in workshops that included over ten thousand women.[24] Exposed early on to French efforts, the British armed forces eagerly took up camouflage; they were later followed, with differing degrees of enthusiasm, by their American, German, and Italian counterparts.[25]

Armies initially deployed camouflage units to reduce the visibility of troops, artillery positions, and forward observation posts that were most likely to be targeted by the enemy. As the war progressed, exercises in concealment were progressively expanded, extending to horses, tanks, trucks, command stations, ammunition dumps, roads, bridges, and railway tracks. With the move toward the aerial bombing of urban centers and industrial sites far removed from the frontline—a practice tentatively initiated in World War I before being mobilized en masse during the next global conflict—it became necessary to also dissimulate, wherever possible, the presence of airfields, ports, shipyards, factories, power plants, refineries, fuel and supply depots, canals, and aqueducts. Misdirecting bombers could also mean concealing or disguising recognizable landmarks, whether natural or man-made, that might not constitute targets but provide useful navigational bearings for enemy pilots. For example, both the British and Germans attempted, with varying success, to mask or reshape existing waterways (the moonlit gleam of which was particularly conspicuous at night) by covering them with boards on piles or pouring coal dust into them.

These latter efforts are examples of a widespread technique of concealment consisting of the use of screens or other barriers to mask the presence or activity of objects. Screening includes exploiting the existing cover provided by natural vegetation or the built environment such as buildings or walls, as well as deploying specially designed materials or structures. Garnished netting and painted canvas sheets, whether draped on or suspended above a position, found extensive use in the Great War. Artillery batteries, machine guns, and tanks were disguised in this manner, but the practice applied further to the concealment of roads, bridges, canals, and railways. While it was not always possible to

FIGURE 39. *French camouflage artists at work, 1915. The figure on the left is the then sexagenarian impressionist painter Jean-Louis Forain, who volunteered for the newly formed camouflage section. Photographer: Jean Corboulin. Photo copyright Paris—Musée de l'Armée, Dist. RMN-Grand Palais / Pascal Segrette.*

FIGURE 40. *Camouflage netting conceals a road from enemy observation in Northern France during World War I. Library of Congress, Prints & Photographs Division, American National Red Cross Collection (LC-DIG-anrc-06254).*

altogether dissimulate such conspicuous transport infrastructure, the enemy's observation of troop movements along it could be successfully obscured (Figure 40). Over the course of the conflict, the British alone reportedly produced around 1 million square yards (840,000 square meters) of canvas, 6 million square yards (5 million square meters) of wire netting, and 7 million square yards (5.8 million square meters) of fish netting.[26] Cost-effective and easy to deploy, netting remains an essential weapon in the armory of visual warfare today. The present U.S. Army version, known as LCSS (Lightweight Camouflage Screen System), is notably able to disguise both the infrared and radar signatures of the object beneath it.

The deliberate release of smoke to cloak troop movements or obscure targeted positions is another common method of screening. During World War I, all the major sides made attempts to cloak the advance and retreat of forces in a chemically manufactured fog of war. In World War II, batteries of carefully positioned smoke generators routinely sought to obstruct aerial views of cities, ports, and industrial sites targeted with bombing.[27] Already heavily reliant on favorable weather conditions for their effectiveness, these smoke screens were rendered largely obsolete by the generalization of the new sensing technology of radar. In response, militaries have since conceived modern obscurants that can absorb or scatter radar pulses, infrared emissions, and even directed laser beams.

A further technique of camouflage in the natural world highlighted by Abbott Thayer, mimicry, strives for "a deceptive resemblance between two kinds of unrelated organisms" rather than a blurring of figure and ground.[28] Various animal species have evolved the means to masquerade as poisonous entities that predators will avoid, or as innocuous twigs, leaves, or flowers that will not draw any unwanted attention. In war, the objective of mimicry is to pass an entity of military significance off as another recognizable but inconsequential object. Under the cover of darkness, First World War combatants became notably adept at replacing bomb-blasted trees with tailor-made imitations, the hollow interiors of which could accommodate an observer or sniper. A whole array of such facsimiles were devised and produced in camouflage workshops throughout the conflict, ranging from fake bushes and haystacks to imitation rocks and even counterfeit cadavers of horses and soldiers.

The First World War dazzle ship is another particular case in the development of military camouflage. Although formally an instance of disruptive patterning in the use of stark color contrasts to confuse an observer as to the proportions and contours of the camouflaged entity, dazzle is uncharacteristic in that it entirely relinquishes any attempt to make its object less visible; indeed, it renders it markedly more conspicuous.[29] Given the highly variable light and weather conditions found at sea, early efforts at disguising ships through combinations and patterns of colored paint proved impractical. With invisibility deemed out of reach, alternative camouflage designs were sought to actively disrupt the enemy's perception and their targeting processes. Allied forces most readily embraced such schemes as a countermeasure to German submarine warfare pursued against their trade routes. Whenever it aimed a torpedo at a ship in its sights, a U-boat was required to "lead" its target—that is, anticipate the position of the ship at the moment the traveling torpedo would reach it. By generating the illusion of a false

FIGURE 41. *HMS* Kilbride *in dazzle camouflage, World War I. Copyright Imperial War Museums (SP 706).*

perspective, dazzle camouflage was intended to confound an observer as to the outline, size, heading, and speed of the masqueraded ship and consequently impede any efforts at tracking its course. Submarine attackers would then take up incorrect firing positions, increasing the likelihood that any torpedo launched might either miss its target altogether or strike a less vulnerable section of the vessel and improve its chances of survival.

An early scheme, advanced by zoologist John Graham Kerr at the outbreak of the war, asserted that it would be quite feasible "to stultify the enemy's range-finders by confusing the details—such as, above all, vertical lines—of which they make use."[30] Inspired by Thayer's writings on animal camouflage, Kerr advocated the "breaking up of continuity of surface and outline by violent color contrasts," with particular attention paid to the masts and funnels that optical range finder operators sought to lock on to for their visual measurements.[31] Having initially dismissed Kerr's suggestions, the Royal Navy eventually turned to a similar plan for disruptive camouflage put forward by painter and naval officer Norman Wilkinson in summer 1917, when it found itself grappling for an urgent response to an intensified German campaign of submarine warfare against Allied shipping. Under Wilkinson's aegis, countless dazzle designs were produced in conformity with his promulgation of a "juxtaposition of violently contrasting colors, black and white predominating, combined in accordance with the laws of perspective."[32] By June 1918, no fewer than 2,300 British merchant and navy ships had been camouflaged in this way, soon emulated by American authorities, with 1,200 American ships receiving the same treatment in the last eight months of the war.[33]

Although artists were intimately involved in establishing camouflage at sea just as on land, its nautical implementation underwent rigorous scientific elaboration according to known laws of optics and thorough experimental testing.[34] Specific ship camouflage designs were extensively assessed through the construction of scale models that were then examined in laboratory conditions that simulated the gamut of lighting conditions at sea as well as views from an enemy submariner's periscope. A 1919 issue of *Scientific American* describes one such theater, located within the research facilities of the Eastman Kodak photography company in upstate New York during World War I (a less sophisticated version of which is visible in Figure 42):

> [A] shallow tank 14 feet in diameter and 12 inches deep was constructed upon which, when filled with water, the model to be examined was floated. Over this tank, a dome of diffusing material was erected which, when lighted from without, gave a lighting condition upon the model similar in quality and distribution to that resulting from the sky illumination. Inside of this dome . . . direct light similar to that from the sun could be thrown on the model, from any desired direction, thus producing the shadow effects of sunlight . . . The models were mounted on a floating support which could be brought into any orientation with respect to the observer, by the use of a cam and motor set in motion at a known speed. They were then examined through a special periscope mounted in a small truck . . . [that] was driven either towards or away from the model at any desired velocity by means of a motor and a speed control cam . . . This made it possible to approach a vessel going at a known speed, and from it make observations at certain intervals to determine the effect of any camouflage pattern.[35]

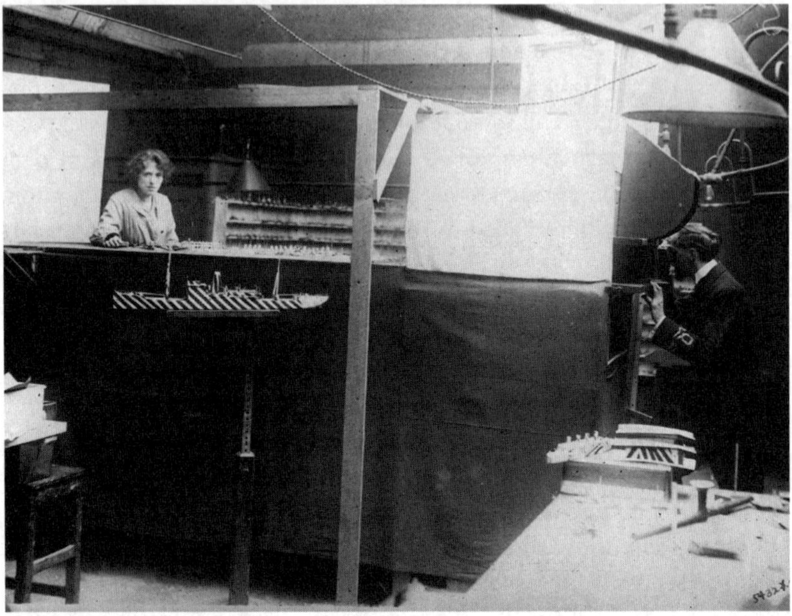

FIGURE 42. *Experimental station for testing naval camouflage designs, Royal Academy of Arts, London, 1919. National Archives and Records Administration (165-WW-70C-9).*

Although postwar assessments of its effectiveness were inconclusive, naval dazzle camouflage did persist into World War II in parallel with renewed investment in visibility-reducing schemes. However, the growth of aerial observation with its distinct perspectival view, and most decisive the deployment of radar and sonar technologies, soon definitively rendered dazzle camouflage obsolete.

While dazzle designs were scientifically evaluated in the manner outlined above, the wild proliferation of camouflage schemes during the two world wars was more often a haphazard affair. The predilections and artistic sensibilities of individual camoufleurs played as much of a role as the methodical application of first principles and systematic empirical assessment of their suitability to specific environments and operational contexts. Throughout this period, most designs were produced by hand, with a corresponding tendency for each application to be idiosyncratic. Camouflage was consequently implemented with variable rigor and effectiveness; at times it proved downright counterproductive when based on impressionistic notions of camouflage, offering little more than psychological comfort to those it was supposed to protect.[36]

If, as gestalt psychologist Wolfgang Metzger observed in 1936, "there is hardly a law of vision that is not found again serving camouflage," then superior military concealment would have to make use of such laws.[37] Acquiring new insights into human perception that could be put to work entailed drawing on the field of psychophysics, or the scientific study of the relation between physical stimuli and the sensations and perceptions evoked by them, and the use of available expertise and investigative methods of physiology, behavioral psychology, and psychometrics. Any findings obtained could then be precisely instantiated through the standardized production of materials and the formulation of strict procedural instructions for their field deployment. The U.S. National Defense Research Committee would lead just such a research program throughout World War II, funding extensive scientific studies of the influence of contrasts in color and brightness on the perception of form, the effects of the atmospheric scattering of light on visibility, or the transient color phenomena induced by changing conditions of illumination in a desert environment. In its postwar assessment, the committee stated, "By studying the inherent limitations of human vision, and by making proper allowance for the effects of the atmosphere, the concealment aspects of camouflage have been reduced, in most cases, to an engineering procedure."[38] Although this remark almost certainly overstates the scientific footing of camouflage at the close of the war, the work undertaken did prefigure subsequent developments that have since submitted the means of concealment to digitized algorithmic treatment and extended their remit to perceptual domains that wholly exceed innate human capabilities.

Camouflage 2.0

In the decades after World War II, camouflage made limited progress, in part because of the notion that advances made by target acquisition technology across the electromagnetic spectrum had rendered traditional methods of concealment irrelevant. The experience of the Vietnam War revived interest after the U.S. military issued elite and special operations forces with disruptive camouflage uniforms in the so-called ERDL pattern, which was created in 1948 but had never previously been deployed. In 1981, the American armed forces adopted a default camouflage uniform for all combat troops, with the chosen Woodland pattern still only a slight alteration of the then thirty-three-year-old ERDL pattern. Evidenced by the continued use of this same uniform until the mid-2000s, the scientific knowledge and technical expertise underlying camouflage design in the visible spectrum remained largely unchanged for the best part of a half century. However, this relative stasis seems to be now giving way to a new period of innovation under the impulse of the progressive digitization of camouflage.

Drawing on experimental work of the late 1970s, digital camouflage uses computer algorithms to produce designs that incorporate multi-scale patterns intended to deceive at a variety of observational ranges.[39] Indeed, disruptive patterns are effective only at certain optimal distances that are typically determined by the expected ranges of tactical engagement. At closer distances, camouflage motifs will be clearly identifiable. At the farthest ranges, their different-hued shapes tend to blur into each other and form a homogeneously colored block that cancels out any disruptive effect and is more likely to stand out against its background environment. The 1981 Woodland Battle Dress Uniform notably expanded by 60 percent the size of the 1948 ERDL pattern in response to the expected increase in the ranges at which camouflage would be tactically relevant in the post-Vietnam era.[40] Contemporary digital camouflage seeks to mitigate, if not wholly overcome, this problem through fractal patterns that remain effective at a wider span of ranges as a result of their properties of self-similarity and scale invariance.[41] Its promoters argue that fractal patterns are pervasive in nature, as displayed by foliage or rock formations, and that the human visual cortex has consequently evolved to treat them as unremarkable and undeserving of further inspection.[42]

Digital camouflage also intertwines Abbott Thayer's modalities of blending and disruption in a more systematic and integrated fashion. The patent for MARPAT, the digital camouflage design acquired by the U.S. Marine Corps in desert and woodland variants in 2004, states that it brings together "principles of human perception, natural camouflage, and psychophysics to create two pattern elements of a macro-pattern and a micro-pattern combined into a single configuration: one to disrupt the features of the subject target, the other to match the subject target to the characteristics of the background."[43] The macropattern hinders recognition of the object's characteristic shape through the arrangement of markings that interrupt its identified axes of symmetry. The micropattern in turn decomposes the macropattern's motif into basic visual units that are arrayed in accordance with the recorded statistical distribution of a designated environment's color components. A further dithering effect commonly used in digital image processing is added through the deliberate injection of visual noise into the pattern, thereby creating the optical illusion of a richer and more natural range of colors than that contained by the actual palette. Controlled trials appear to support the claims made for the superior efficiency of digital camouflage, showing the detection time for targets bearing the MARPAT pattern to be 2.5 times longer than for NATO single-scale camouflage, with a further 20 percent increase in the required recognition time after detection.[44]

Over and above these technical details, the key distinguishing feature of digital camouflage is its computerization of the design process, weaving together into an algorithmic mesh the various abstractive strands spun from known optical phenomena, identified psychophysical characteristics, and the expected operational environment.[45] Despite its widespread misconception, the common association of digital camouflage with the distinctive pixelated appearance of many of its patterns is in fact a trivial and nonessential relationship. Indeed, it is quite possible to generate multiscale fractal motifs from constituent elements of any shape, quadrilaterals merely being at present the easiest way to computationally render such designs.[46] This misapprehension accounts for the much-publicized debacle surrounding the U.S. Army's 2004 introduction of its Universal Camouflage Pattern (UCP) for all combat uniforms. For one, the pattern's purported suitability for concealment in every environment resulted in compromises that rendered it glaringly ineffective in all of them. Moreover, its particular pixelated look appears to have largely been an aesthetic choice guided by a poor understanding of the underlying principles of camouflage, along with insufficient testing. Faced with mounting criticism of its shortfalls, the Army was eventually forced to withdraw the $5 billion UCP from field operations before announcing in 2014 its wholesale replacement by a variant of an older digitally designed (but neither pixelated nor fractal) pattern.[47]

The UCP fiasco emphasized that the potentialities and technical requisites of digital camouflage are still not fully understood, but it also highlighted that the military institutions' choices of camouflage patterns cannot be explained solely by reference to their powers of concealment. However otherwise unobtrusive, national uniforms must remain distinctive enough to allow soldiers to promptly recognize their own in the field and avoid potentially lethal cases of mistaken identity. For this very reason, militaries have previously rejected camouflage patterns because they too closely resembled ones adopted by opposing forces. In a similar vein, the U.S. Army mandated in 2004 that infrared patches should be affixed to uniforms whose design purposefully minimized their visibility in the infrared spectrum, thereby reducing the risk of friendly fire incidents during nocturnal operations. Although modern militaries have abandoned the showy uniforms of the nineteenth century, their professional garb nevertheless continues to be an important source of institutional pride and an emblematic marker of collective identity. Aesthetic considerations, at a remove from the unadorned functionalist dictates of camouflage, inevitably inform sartorial decisions. Anthony King has, for instance, persuasively argued that the U.S. Army's initial enthusiasm for adopting pixelated camouflage was due in no small part to its perceived symbolism of the wider digitization of the armed forces. Pixelated motifs, all the more in impractical "universal" color schemes, may have less to do with enhancing dissimulation than with visually signaling to the world the institution's association with cutting-edge technology.[48]

While the design of uniforms cannot be entirely reduced to bare functionalism, the advent of digital camouflage nonetheless clearly marks a further stage in the science of concealment. Whereas early camouflage was chiefly the handiwork of visual artists and their particular interpretations of a few broad principles, its production is now increasingly being directed by finely tuned algorithms that meticulously exploit the neurological and computational biases and blind spots of the perceptive organs scanning the battlespace. These do not deliver any definitive panacea to the problem of dissimulation, however. One study suggests, for instance, that fractal camouflage patterns could be particularly vulnerable to the judicious addition of stochastic noise to sensor images, exposing by this means the subtle regularities of their motifs.[49] The game of cat and mouse that has always characterized camouflage's entanglement with surveillant countermeasures will persist but is ever more likely to pass through informational processes and algorithmic contests that bypass the biological thresholds of human perception and cognition.

There is perhaps no better evidence of the human organism's decentering than the advent of stealth technologies responding to the giant leaps realized in the detection and recognition of targets across the electromagnetic spectrum. So-called multispectral camouflage seeks to simultaneously shield entities from the perceptual clutches of optical, infrared, radar, and laser sensors (along with other nonelectromagnetic sensors operating in the acoustic or chemical domains). Accordingly, camouflage has become increasingly understood as an exercise in signature management, whereby a given target's signature corresponds to its characteristic aggregate of distinctive signal features across the array of relevant sensorial fields. As the martial gaze's perceptual reach has expanded, the methods of concealment and obfuscation have become ever more sophisticated and onerous. Following on from the development of infrared suppressants and the meticulous shaping of radar-evading aircraft, the latest endeavors include attempts at engineering materials down to the very atomic scale of matter and producing dynamic camouflage systems that can automatically adjust to their immediate surroundings.

False-color infrared film designed to distinguish camouflaged equipment from surrounding vegetation was introduced in the 1940s. It was soon countered by the application of infrared-reflecting paint. Similarly, the 1970s-era adoption of fabric dyes that can absorb infrared light was a response to the diffusion of image intensifiers and concomitant night-vision capabilities. As specified by the U.S. Army's latest relevant field manual, "standard-issue camouflage materials" should all "exhibit an artificial chlorophyll response" at selected infrared wavelengths. Put more simply, they should appear as natural vegetation across the widest possible spectrum of light.[50] In the case of thermal imaging and heat-seeking sensors, countermeasures suppress and diffuse thermal radiation generated by vehicles and personnel, albeit so far with only mixed results.

Against radar, any stealth initiative seeks to minimize an object's radar cross section: the amount of electromagnetic energy reflected back from a radar pulse. While a target's size normally correlates with its reflectivity to radar, radar-absorbent paint or materials and the

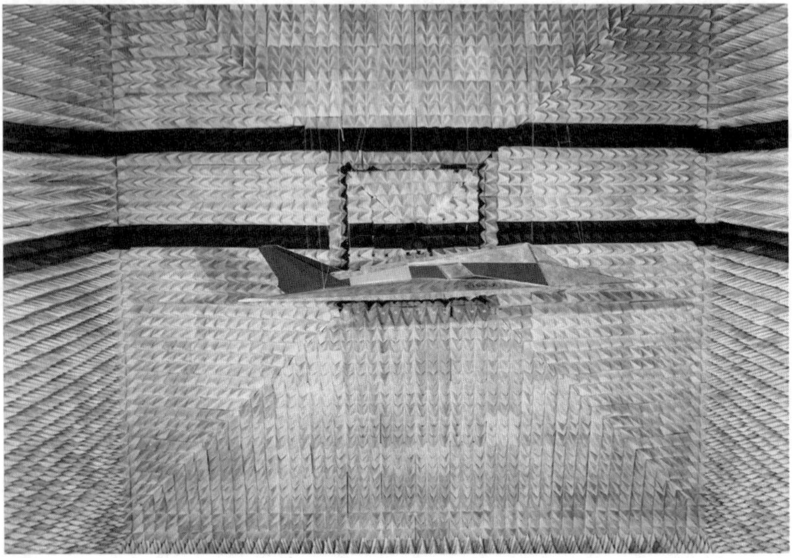

FIGURE 43. *F-117 model in an anechoic chamber for the measurement of its radar cross section. San Diego Air and Space Museum Archive (10_0016118).*

careful shaping of a body's surfaces and angles can achieve significant reductions. As the first operational stealth aircraft originally conceived in the 1970s, the F-117 Nighthawk traded aerodynamics for a reduced radar cross section, its characteristic boxy outline resulting from the adoption of flat surfaces for the deflection of radar signals (Figure 43). By the following decade, increased computing power led to advances in the modeling of radar returns as well as design and manufacturing, enabling the curved surfaces of the B-2 bomber's flat flying wing design. A major leap in aircraft stealth technology, with a unit cost of around a billion dollars, the B-2 reputedly has the same radar cross section as a small bird and a thousandth of that of the B-52 bomber with a similar wingspan.[51] Building on these aeronautical developments, the design of modern warships such as destroyers or frigates also increasingly seeks to maintain a low radar profile. Finally, radar plays a major role in the detection and identification of infantry forces, causing military organizations to strip radar-reflecting metal, wherever possible, from helmets, body armor, buttons, buckles, and water canteens.[52]

While stealth technologies have successfully reduced the electromagnetic signatures of target entities, the perceptual apparatuses arrayed against them have not simultaneously stood still. Advances in signal processing are continually widening military radar's frequency bands beyond those which stealth aircraft have been optimized to evade.[53] Forthcoming passive radar systems, which listen for the echoes from ambient radio emissions without emitting any signals of their own, have the potential to negate existing stealth technologies in addition to being much harder to locate and neutralize.[54] A major review of American national security in 1999 concluded that "physics probably favors detection and the ultimate demise of stealthy systems," which is reinforced by the highly adverse cost ratio that characterizes stealth development relative to that of new sensor technologies.[55]

Nevertheless, recent innovations in materials science have opened up new stealth possibilities that may still arrest, or at least slow, such a trend. The emerging field of metamaterials, which holds particular promise, is engineering tiny nanostructures that can channel a range of electromagnetic frequencies through and around them, suppressing the scattering of energy and varying its refraction. In principle, such structures could cloak a target, rendering it wholly imperceptible (or at least significantly dampening its electromagnetic signature) as well as deflect laser weapons operating at the same frequencies. Laboratory experiments, financed by support from military agencies, have already produced electromagnetic cloaks that shield very small objects from certain microwave and near-infrared frequencies.[56] Although practical applications are not expected to materialize for decades, if at all, metamaterials certainly have the potential to usher in a new revolution in military stealth technology.[57]

A different line of research seeks to remedy one of the fundamental limitations of traditional camouflage, which can only be optimized for a very specific location or on the basis of a statistical aggregation of its envisioned surroundings (as seen in the common variants of generic woodland, jungle, desert, and arctic patterns). Any entity moving through a variety of diverse backgrounds is vulnerable to marked discrepancies in the effectiveness of its camouflage that at times may even become positively counterproductive. The heterogeneous urban environments where military operations are increasingly taking place are particularly challenging in this regard. An active system of camouflage able to dynamically adapt its color, shape, or brightness to match its immediate background constitutes an evident but technically exigent solution to this problem.

In the natural world, organisms equipped with active camouflage, such as the chameleon and various cephalopods, use chromatophore skin cells that rapidly alter their pigmentation to mimic their surroundings. Unsurprisingly, research into these biological mechanisms is a major source of inspiration for the elaboration of artificial devices.[58] These speculative systems of active camouflage typically involve wearable displays (such as smart textiles or flexible video screens) that are used to generate images on the fly. While display surfaces alternating between predetermined camouflage patterns are already possible, more ambitious schemes strive for the wearer's total invisibility by displaying a semblance of their exact background as recorded by an adjunct sensor. However, existing arrangements are highly restrictive because they typically require precise knowledge of the observer's position (and therefore of their perspectival viewpoint) to sustain the illusion.[59] Overcoming this constraint would likely necessitate displays with multiple viewing angles and the construction of three-dimensional digital maps of the environment. For now at least, the technical obstacles remain considerable.

One of the earliest attempts at active camouflage can be traced back to World War II and the American antisubmarine campaign in the Atlantic Ocean. The U.S. military was concerned with reducing the visibility of their incoming aircraft and ensuring that an enemy submarine would not have the time to submerge and escape the dropping of depth charges. Existing methods of camouflage proved ineffectual because even an aircraft painted white will typically appear as a black silhouette against the sky. Considerable effort consequently went into investigating the proposal that lights distributed along the fuselage and leading edge of the wings could, if set at the right intensity, render the aircraft indistinguishable from the background sky from the target submarine's perspective. Tests soon showed that the so-called Yehudi lights could indeed decrease the visible range of a bomber from 12 miles (19 kilometers) to less than 2 miles (3 kilometers), but the technology was never deployed because of the contemporaneous development of radar. Crucially for our discussion, the system was designed as a form of adaptive camouflage, with the intensity of the lights matched to the brightness of the background sky as measured by a photoelectric cell. The necessary adjustments in luminosity were made either manually or by the operation of an automatic control mechanism.[60]

BAE Systems' work on its ADAPTIV camouflage system is a modern-day incarnation of Yehudi lights for the infrared range of the electromagnetic spectrum. Intended to confound heat sensors, ADAPTIV uses a honeycomb of temperature-controlled hexagonal panels that can be fitted to any vehicle. The panels are designed to rapidly adjust their temperature to match that of their surroundings as measured by onboard thermal sensors, thereby concealing the target's expected heat signature. BAE Systems further claims that the temperature of the individual modules (effectively the single pixels of a heat image displayed by them) can be set so as to create patterns that "make a tank look like other objects, such as a cow or a car, or bushes and rocks."[61]

Camouflage appears set on a trajectory shadowing that of the wider technology of imaging in shifting from static representations to dynamic displays updated synchronously with the sensed environments in which they seek to disappear. Whether active camouflage, metamaterials, or algorithmic designs can tip the balance back in favor of imperceptibility remains to be seen—or perhaps, more accurately, unseen. The military efforts expended on such emerging techniques are certain to only intensify as new technologies progressively erode camouflage's efficacy and as remote targeting renders exposed entities ever more vulnerable. However, there remains one more weapon in the arsenal of concealment for us to consider. Paradoxically, it is one that does not evade the martial gaze but rather purposefully draws its attention and potential wrath.

What You See Is Not What You Get

If camouflage conceals a military target's presence, shape, or identity, then decoying inversely displays their simulacrum. While appearing antithetical, both practices share in common the objective of masquerading reality to the enemy and draw on the same repositories of knowledge and experience. Generally speaking, any entity liable to being camouflaged is susceptible to being decoyed. In practice, decoys serve several distinct, if nonexclusive, purposes. They complement camouflage efforts in drawing the enemy's attention to a counterfeit entity and away from any genuine target nearby. Decoys can also expend and misdirect enemy resources dedicated to observation and targeting, notably by provoking an attack on them and providing valuable information on the adversary's positions or tactics. Finally, they can deceive the enemy over one's strength and intentions by creating the illusion of a larger military force than exists, masking the redeployment of actual troops and equipment, or affecting to undertake certain courses of action so as to shield the real plan from view.

The intense camouflage efforts of World War I were matched by a commensurate creation of varyingly successful decoys that included soldier mannequins, plywood silhouettes, dummy heads, counterfeit artillery, mock vehicles, fake trenches, phony installations, and bogus landmarks. In 1917, a story in a French satirical publication derided the enthusiastic embrace of such visual deceptions. A fictitious camoufleur presented there his impassioned proposal to "invert the landscape" under the cover of darkness such that upon sunrise, the Germans would mistake their own lines for those of the enemy, attack their own reserves, and march back toward the Rhine. The tale concludes with its hapless protagonist being led to the madhouse, still promulgating his "war-winning" stratagem to all who will listen.[62] In the very same year, the French military hatched an earnest, but nearly as outlandish, plan for the erection of a replica of Paris in the surroundings of the capital so as to misdirect incoming German bombing raids at night. For its realization, the scheme called for counterfeit landmarks, avenues, factories, railways, and train stations to simulate the city's subdued wartime illumination. The French eventually abandoned the construction of the decoy Paris at the conclusion of war, but not before assembling a likeness of the Gare de l'Est.[63]

However, the aborted simulacrum of Paris did prefigure how decoys would be arrayed against the foremost threat in the next war, namely air power and its surveillant arm. After the fall of France in 1940, Britain faced intense bombing by the Luftwaffe so as to compel a negotiated peace with the Reich. Because the opening nightly air raids' leading objective was to neutralize the RAF, the British primarily channeled their initial decoying efforts into setting up dummy aircraft factories and aerodromes through the judicious positioning of illuminated buildings and the simulation of runways by flare paths. When the Germans decided

FIGURE 44. *Australian soldiers of the Fourth Field Company of Engineers carrying a dummy tank with camouflage markings, 1918. Copyright Imperial War Museums (E AUS 4938).*

to extend operations to the daytime, the British were forced to supplement these original "Q" sites by more elaborate "K" sites that had to be convincing in the light of day. Similarly, the Blitz's expansion of bombing to industrial and population centers would motivate the British to simulate urban lighting and create diversionary fires to mislead Luftwaffe crews. In total, over five hundred such sites were established during the Battle of Britain, with the ersatz RAF aerodromes performing particularly well in reportedly drawing as many raids as the real fields.[64]

As the tide of the war turned and the Allies unleashed a relentless campaign of strategic bombing, the Axis powers were in turn compelled to manufacture and deploy decoys that might lure aerial attacks away from their true objectives. Echoing World War I plans for a fake Paris, one German endeavor visually shifted targets away from Hamburg's center by laying boarding over 200,000 square meters of open water and painting onto it a semblance of building blocks and streets as viewed from the air, putting up a counterfeit bridge, and constructing a replica train station with tracks and carriages.[65] As in the Battle of Britain, the struggle for aerial supremacy made aeronautical assets a prime target of bombing raids and thus a prominent focus for the use of decoys. According to one estimate, by 1945, no less than a quarter of the planes visible on Japanese airfields were dummies.[66]

Decoys that could durably fool intelligence analysts, who meticulously studied aerial photographs, were difficult for armies to produce. Moreover, poorly executed enterprises risked being counterproductive by highlighting to the enemy the likely proximity of camouflaged assets deemed valuable enough to warrant their imitation. Inexpensive vehicle decoys, in the guise of painted ground surfaces or cut-out silhouettes, were far less likely to stand up to scrutiny compared to three-dimensional reconstructions made of canvas over a wood framework or inflatable rubber (or even simply vehicles retired from service).[67] Sustained deception also typically required the continuous simulation of activity: the movement of logistical supplies, the presence of fresh vehicle tracks, and other expected signs of maintenance and live operations. An overly conspicuous decoy tended to raise immediate suspicions, leading to military forces deliberately applying sloppy camouflage to give the illusion of attempted concealment. Once attacked by the enemy, a decoy site often demanded the addition of fake bomb damage if the subterfuge was to endure. While the standard required to deceive a trained photographic interpreter could therefore not always be met, a decoy might still prove worthwhile if it were able to mislead or confuse a bomber crew under the time-sensitive pressure of a live mission.

For all this, forces could still orchestrate large-scale deceptions of strategic import through careful planning and scrupulous execution. The Allies notably achieved victory in Egypt in no small part due to elaborate deceptions undertaken in preparation for the Second Battle of El Alamein. By successfully disguising tanks as trucks and deploying hundreds of decoy vehicles and artillery pieces, Montgomery's Eighth Army gained a decisive element of surprise over the Axis forces.[68] For their part, the Twenty-Third Headquarters Special Troops unit of the U.S. Army, best known as the Ghost Army, used a combination of visual, sonic, and radio ruses to impersonate entire divisions across France and Germany in the final year of the war.[69] Most impressive of all was Operation Bodyguard, a strategic deception plan that drew on the full panoply of camouflage, decoying, and disinformation techniques to dupe the Germans about the true time and place chosen for the invasion of northwest Europe. Throughout the spring of 1944, the British succeeded in persuading German intelligence of the existence of a Fourth Army of 250,000 men and hundreds of tanks, armored vehicles, and artillery pieces amassing on the Scottish east coast in preparation for an invasion of Norway, thus preventing the allocation of further reinforcements to defend against an Allied crossing of the English Channel.[70] The concoction of another fictitious army group, supposedly under the command of General Patton, similarly convinced the Wehrmacht's leadership that the main offensive into France would take place in Pas de Calais, an erroneous belief sustained for weeks after the actual Normandy landings had taken place. The Soviets, who themselves had been caught unprepared by the German aggression of June 1941, also eventually perpetrated large-scale deceptions on the Eastern front that played a major role in the pivotal victory of the Battle of Stalingrad.[71]

Just as military perception has extended across the electromagnetic spectrum, decoying has naturally followed the same course in its forging of counterfeit target signatures. An early response to radar technology was the corner reflector, a simple passive device whose perpendicular arrangements of metallic surfaces sends back strong radar echoes that pass for much larger structures.[72] Just as conventional decoys can be used to draw enemy fire and reveal its point of origin, electromagnetic lures such as decoy missiles can serve to disclose the location of radar antennae activated by their presence, thereby exposing them to subsequent attack. Certainly any credible visual decoy must today simultaneously mimic the corresponding infrared and radar target signatures if it is not to be unmasked in short order.

Although not its exclusive preserve, electromagnetic decoying has given rise to common practices of bedazzlement that generate signal clutter or noise, which overwhelms sensors and conceals relevant signals. Referencing the use of powerful searchlights to blind enemy soldiers in the Russo–Japanese war of 1904, Friedrich Kittler noted the advent of "an actively armed eye that no longer simply optimized its own perception, like the telescope and the microscope, but also reduced the perception of the enemy to zero."[73] During World War I, a similar principle could be found in the German practice of frustrating the Allied sound-ranging efforts by spreading out their artillery battery positions and firing them simultaneously so as to prevent individual detonations from being distinguished from one another.[74]

FIGURE 45. *Royal Air Force bomber releases window during a daylight raid over Germany, 1944. Copyright Imperial War Museums (CL 1405).*

Notwithstanding these noteworthy antecedents, the electromagnetic realm has made bedazzlement an essential weapon in the arsenal of perceptual countermeasures. For instance, the use of chaff involves releasing dense clouds of small metallic fibers tailored to maximize their reflectivity of radar signals. First deployed in the shape of thin strips of aluminium (called "window") by the British in World War II, chaff can serve to either generate radar echoes that resemble legitimate targets ("spoofing") or simply saturate the perceptual field with signals that drown out the actual presence of real forces (Figure 45). Electronic forms of jamming entail the purposeful emission of radio signals, which serves to overwhelm radar receivers altogether or to more subtly misdirect their interpreters. Similarly, powerful radio transmitters targeted the navigational radio beams used during World War II to guide bombers to their targets, such as Gee and its German counterpart, Knickebein, deflecting the path of the beams and leading aircraft astray. Today, the contemporary navigational technology GPS is not immune to the threat of jamming and spoofing, prompting militaries to adopt signal encryption and backup guidance systems in response.[75] A pyrotechnic pendant to chaff, the flare's intense heat discharge draws infrared sensors to it, particularly those guiding missiles to a vehicle's thermal emissions.[76] Finally, a more recent counter to infrared guidance is electro-optical dazzling, where a blinding laser beam is directed straight at the sensor's eye, potentially disabling both laser-guided and heat-seeking munitions.

How to Disappear Completely

For all the ingenuity and industry expended in developing technical ripostes to the martial gaze, the laws of physics and unrelenting trajectory of increasingly precise sensors will likely present an ever greater hurdle to concealment. As the sensorial mesh tightens over the battlespace, Martin Libicki asserts that "even with stealth, everything ultimately can be found" because "sensors of certain minimum discrimination placed close enough together can, at some epsilon, catch anything."[77] In practice, the likelihood of consistently defeating every technology of dissimulation remains remote, especially with a deployment of the full gamut of decoys and offensive kinetic and electronic measures for degrading and misleading military perception. However, the ever-heightening powers of sensorial discrimination do threaten to restrict the most advanced techniques of concealment to just a few onerous systems that only the best-resourced forces can deploy in limited numbers. The U.S. military has only been able to afford the manufacture of twenty-one B-2 stealth bombers, and its scheduled B-21 successor will probably not prove less costly. In this light, we need to take stock of the increasing prevalence of different kinds of responses to the martial gaze from those we have surveyed so far. These countermeasures do not primarily rest on expensive cutting-edge technologies, and leading militaries possess no necessary monopoly over them. The considerable diversity of such efforts can be broadly encapsulated as strategies of generalized concealment, dispersal, and expendability.

A first stratagem purposefully displaces the arena of conflict to physical terrains that hamper the logistics of perception and its correlated processes of targeting. The pursuit of favorable geography for military advantage is hardly new, of course, having long been recognized as an elementary component of tactical and strategic acumen capable of decisively offsetting large imbalances in contesting forces. Yet in the age of lethal visibility, combatants who expose themselves in open terrain are vulnerable as never before, as the Iraqi army discovered at its considerable expense in 1991. Writing a few years later, military analyst Ralph Peters concluded that "battlefield awareness may prove so complete, and precision weapons so widely-available and effective, that enemy ground-based combat systems will not be able to survive on the deserts, plains, and fields that have seen so many of history's great battles."[78] In contrast, the density and obstructions of mountainous, forested, and jungle milieus severely impede the range and accuracy of modern sensors and considerably restrict the space in which weapon systems can operate. These environments' abundance of cover greatly assists belligerents in both dissimulating their presence and sheltering themselves from attack. Repeatedly adopted by insurgencies and guerrillas facing larger and better equipped forces, natural terrain's ability to negate much of the latter's perceptual dominance of the battlespace has been amply demonstrated.[79] Increasingly, though, the urban environment, with its high concentration of buildings and civilian populations, is also leveling the playing field through the "clutter of concealment."[80] Given the frantic pace of planetary urbanization and intensifying global connectivity, the countryside may well be ceding to the city the privileged site of asymmetric warfare.[81] Finally, another analogous approach is the descent into the underground. Although its military utility reaches back millennia, the subterranean realm has offered a particularly appealing refuge from the ratcheting deadliness of modern war.[82] As Paul Virilio notes, the architectural figure of the concrete bunker manifests the progressive ensconcing of defensive structures into the landscape, so as "to hide from view and to avoid shocks."[83] Underground tunnels and complexes featured significantly in the two world wars, with Germany and Japan even attempting to relocate some key war industries in the latter conflict's final stages. Vast subterranean facilities have only further proliferated since, burrowing ever deeper into the earth's crust.[84] In addition to the burying of national security sites for protection against nuclear attack, extensive networks of tunnels for the hosting and movement of personnel and matériel has become a recurrent feature of armed conflicts, including those in Korea, Vietnam, Afghanistan, Gaza, and Lebanon.

The principles of dispersal and expendability characterize a second type of response. The American defense establishment has entertained arguments promoting their merits for some time now as it confronts the likely consequences following from the inevitable diffusion of their advanced perceptual technologies and precision munitions. Advocates point to the growing vulnerability of the large expensive platforms that have been traditionally favored by the military and its affiliated congressional–industrial complex, calling for a concerted shift to smaller and cheaper systems that could be deployed in much greater numbers.[85] In particular, advances in computing and robotics offer the prospect of saturating the future battlespace with a multitude of disposable systems that could fulfill any of the functions of sensor, information processor, telecommunication relay, weapon, or decoy. A 2014 report authored by the then incoming U.S. Deputy Secretary of Defense submits:

> Instead of building ever-smaller numbers of exquisite crewed platforms to penetrate an enemy's battle networks, large quantities of low-cost, expendable unmanned systems can be produced to allow U.S. forces to overwhelm enemy defense with favorable cost-exchange ratios. By relying on smaller size and superior numbers, unmanned systems can be built to be lost in combat, making survivability a characteristic not of any individual platform but of a swarm of systems, operating together.[86]

The reference to a "swarm of systems" relates to the tactic of swarming that is typically associated with calls for military dispersal and decentralization of command. Although already practiced in the past by groups such as Genghis Khan's Mongols, contemporary proponents of swarming point to the diffusion of information processing and transmission technologies by which distributed units can achieve unprecedented levels of adaptive cooperation.[87] In this way, units can remain scattered across the battlespace, minimizing their target signature and ready to rapidly coalesce against a target for a coordinated pulse attack with a concentration of force that would have previously only been possible through the conspicuous massing of forces.[88] With the growth of autonomous systems, some analysts envisage the constitution of a "reconnaissance-strike swarm" composed of billions of "tiny, insect-like drones" interacting with each other to produce an emergent collective intelligence whose responsiveness on the theater of operations will outpace that of any human overseers.[89]

Swarming is, however, far from being the sole preserve of advanced militaries. Nonstate groups are already adopting decentralized tactics and coordinating their violent actions through commercially available telecommunication devices. Attacks in Mumbai (2008) and Paris (2015) showcased the deadliness of this approach and the difficulties it poses to law enforcement in mounting timely responses to these operations. Jihadist groups have also turned expendability into a significant asset, not by resorting to unmanned systems (yet) but through the use of suicide attacks that enact a covert weaponization of the human body. In the words of Achille Mbembe:

> The "suicide bomber" wears no ordinary soldier's uniform and displays no weapon . . . The candidate for martyrdom transforms his or her body into a mask that hides the soon-to-be-detonated weapon. Unlike the tank or the missile that is clearly visible, the weapon carried in the shape of the body is invisible. Thus concealed, it forms part of the body . . . The body is transformed into a weapon, not in a metaphorical sense but in the truly ballistic sense.[90]

But is the human time bomb's covert weaponization not simply the purest tactical expression of a broader vector of strategic concealment that sees armed combatants dissimulating their presence at the very heart of their enemies' societies? Philosopher Reza Negarestani labels this "dissolution into the enemy" as "hypercamouflage," "a silent and fluid military infiltration" of civil society by antagonists who endeavor to veil their intentions until the very moment of their assault.[91] Hypercamouflage thus participates in a process of "endo-militarization of peace" that dissolves any notion of the battlefield as a delineated space, collapsing the respective categories of war and peace and spheres of military and civilian. In turn, the state's response in placing its entire citizenry under ever-greater suspicion only enacts its own "retro-militarization" in which its "'teeth' flow back down the 'tail' in a process without obvious terminus (short of the fanging-up of the entire social body)."[92]

Hypercamouflage can clearly be construed as a coherent asymmetric strategy responding to the domination of the martial gaze.[93] However, the latter has itself contributed to blurring the contours of war and peace by enabling practices of global targeting that increasingly pay little regard to whether they take place within formal theaters of war. Although a phenomenon that can be traced back through the twentieth century and the associated totalization of war, the disappearance of the battlefield, as both a bounded geographical space and regulated legal space, has become particularly pronounced under the war on terror's seemingly indeterminate, planetary remit.[94] In proportion to this twilight of the battlefield, the norms that have customarily regulated conduct on it, however imperfectly, have weakened. The purely military advantages that could be gained through practices of remote targeting or concealment were in the past restrained in some measure by the wider social construction of the battlefield and its related discrete spatiotemporalities of war and peace. Those fetters appear to be decisively loosening today, in however piecemeal and haphazard a fashion.

For instance, the various efforts at invisibility previously discussed have been tempered throughout by one notable restriction placed on the military uniform, namely the requirement that it unambiguously identify its wearer as a member of recognized armed forces.[95] Indeed, the international laws of war mandate that combatants engaged in armed conflicts must bear distinctive uniforms that clearly differentiate them from both their military adversaries and the civilian populations that should be immune from targeting.[96] While camouflage, decoys, and willful misinformation are permitted, the resort to such deceptive measures as the wholesale removal of uniforms or the impersonation of enemy uniforms is prohibited. These constraints have at different times and places been disregarded, or at least relaxed, particularly in the context of intelligence collection and Special Forces operations.[97] The increasing reliance on these types of operations since the inception of the war on terror has, however, certainly contributed to the erosion of these norms. Already during the 2001 invasion of Afghanistan, U.S. Special Forces operatives dissimulated their presence amid the local allied forces by adopting indigenous attire, customary beards, and native mannerisms. Pentagon lawyers insisted that there was no breach of international law, as the operatives were professedly wearing nonstandard uniforms that were sufficiently distinctive and consistently displayed to mark them out as members of a discrete combatant group, but there is no doubt that some willful obscuring of their identity did take place. Certainly Taliban fighters have for their part periodically led surprise attacks in which they have donned U.S. Army, Afghan National Army, or police uniforms to disguise themselves.[98]

Elsewhere, the Russian annexation of Crimea and subsequent incursions into Eastern Ukraine in 2014 were marked by the sudden appearance of "little green men," armed soldiers in balaclavas and camouflage uniforms bearing no military insignia and claiming to be members of spontaneously formed local militias, if they identified themselves at all. The Kremlin steadfastly rejected allegations that these were its own forces, adamant that it had no military personnel on the ground before eventually conceding that some had indeed been deployed. Among the other suspected ruses were military vehicles adorned with peacekeeping insignias and the border crossing of large "humanitarian aid convoys" made of army trucks hastily painted white and thought to be supplying weapons and other material support to separatists (although they may in fact have merely been a decoying measure drawing attention away from more discreet border crossings). These maneuvers seem to have primarily served the strategic purpose of flummoxing Russia's adversaries and hindering the formulation of a cohesive response to its actions, rather than acting as a more tactical intention to elude targeting. A line can also be traced from these subterfuges to the long-standing Russian military doctrine of combined deception, concealment, and disinformation called *maskirovka* (something masked), as practiced throughout the Soviet era with antecedents stretching back into imperial times.[99] Nevertheless, this systematic divestment or blurring of military identifiers can still be seen as a further symptom of the evanescence of the bounded space of the battlefield and the norms associated with it.

The American policy of targeted killings, under which the military has launched drone strikes in Afghanistan, Pakistan, Yemen, Somalia, and beyond, is probably the clearest illustration of contemporary targeting's global reach and its corrosive effects on the conventionally bounded framework of war. Surveyed by the attentive eyes of drone aircraft remotely piloted from across the world, individuals whose precise identities are most frequently unknown but who are nevertheless deemed to manifest dangerous "patterns of life," are marked for elimination following the operative logic of preemption.[100] Through this "individualization of targeting," the United States has been moving toward redefining the legitimate space of war as wherever a designated target finds itself, inverting the customary precedence of the former.[101] As for the prospective targets, generally labeled as unlawful combatants for failing to meet the necessary criteria of international law—inter alia because they are not judged to be wearing uniforms—the persistent presence of an invulnerable lethal eye can only drive them to ever more radical forms of concealment. For Grégoire Chamayou, the exercise of armed force thus comes to align itself more closely with the practice of the hunt than of war through a "hostile relationship" that, in the words of one military source, comes to resemble "a competition between hiders and finders."[102] No longer characterized by the interchangeability of these roles typical of the major conflicts of the last century, the hunt returns us to the very origins of camouflage, resting on a structural asymmetry that installs advanced militaries in the perpetual role of pursuer. However, as the terrain of the hunt extends to encompass the globe's every nook and cranny, so must the martial gaze inexorably turn itself on the very societies from which it arose.

CONCLUSION

A Global Imperium of Targeting

> The gaze that sees is a gaze that dominates; and although it also knows how to submit itself, it dominates its masters.
>
> —MICHEL FOUCAULT

Over the course of the modern era, the ascendancy of a regime of global targeting has made perceptibility synonymous with a fatal vulnerability. In establishing a rational and systematic procedure for the visual recording of spatial measurements, the Renaissance's invention of linear perspective laid the foundations of a new architecture of control. Conjugating the fields of optical science, pictorial representation, and land surveying through a single geometrical method, perspective set in motion a conjoined rationalization of vision and mathematization of space, thereafter implemented within ever more sophisticated and efficacious sociotechnical assemblages. Decisively accelerated and oriented by the exigencies of armed conflict, above all through the industrial and total forms taken by war in the twentieth century, the constitutive activities of sensing, imaging, and mapping have combined to bring perception and annihilation into ever closer alignment.

A martial gaze roams our planet, from the watchful satellites peering down from their silent orbits to the infrasonic sensors probing its subterranean and subaquatic depths. Our contemporary condition is rapidly converging on a globalized targeting capable of tracking and eliminating an entity wherever it is found. Moreover, this planetwide targeting is increasingly granular and individualized. It no longer merely distinguishes targets by their displayed membership of a hostile armed force within a recognized theater of war. Instead, specific human bodies are routinely singled out for elimination by reference to their known individual identities or their recorded patterns of behavior and interpersonal associations.[1] Yet as targeting takes on the character of a global hunt, the concealment and dispersal of its prospective prey deepens accordingly. In the twenty-first century, the dialectical play of these tendencies is rapidly dissolving the spatiotemporal and normative constraints of the conventional battlefield. Untethered from discrete spheres of armed conflict and committed to relentlessly pursuing individual antagonists, the martial gaze must inevitably cast itself on the very civilian society that it was ostensibly erected to defend from external threat.

The Gaze Turns Inward

Although they have hardly been consistently or universally upheld, distinctions between states of war and peace, domestic and international arenas, and military and civilian spheres have been central to the Westphalian international order and its Weberian nation-states.[2] These delineations collapse when targeting is globalized and war as a discrete and bounded manifestation of armed struggle gives way to temporally discontinuous and geographically disparate "states of violence" tantamount to a "strategic shattering."[3] The hallmark of this new modality of conflict is the abrupt, unannounced eruptions of violence that dislocate the routine of civilian life in a specific locality while simultaneously embedded in a dense global meshwork of causes and effects. Derek Gregory has underlined this "emergent 'event-ful' quality of military, paramilitary and terrorist violence that can, in principle, occur anywhere."[4] Such sudden lacerations can be caused by the remote control of an advanced weapon system from an air-conditioned trailer on the other side of the planet or by the close-up detonation of a transnationally activated, hypercamouflaged human bomb in a crowded urban space.

The planetary expansion of militarized perception and targeting has not secured the societies at the forefront of their deployment. Instead, it has actively participated in a molecularization of war manifest in the centrifugal dispersal of the warring activity and the corresponding proliferation of the logic of security throughout the social body. Molecular war dispenses with opening declarations of hostilities or armistice agreements, recognizes no fronts or rears, pays no heed to territorial advances and retreats, and forgoes sustained contiguous operations. Directed less by overarching strategic rationales than their own self-perpetuating diagrams of operation, punctuated acts of violence momentarily conjoin distant points on the globe through the circulatory logistical flows of information, capital, and matériel. The Obama administration's early decision to rebrand the "global war on terror" as "overseas contingency operations" seems here particularly revealing in its acknowledgment of the discontinuous, ad hoc, and fundamentally open-ended nature of America's global projections of force.[5] Yet the bracketing of these actions to overseas territories obfuscates the swift erosion of distinctions between domestic and foreign spaces taking place today.

The total wars of the twentieth century overtly militarized entire societies, concentrating their energies toward a decisive victory over an identified enemy. Molecular war offers instead fitful discharges of violence as its interminable horizon, thereby inducing a subtler yet potentially no less comprehensive process of domestic securitization. Following from the unremitting injunction to protect civilian populations and safeguard the smooth logistical workings of postindustrial societies, the logic of security pervades the everyday through the constant anticipation of danger and indefinite management of risk.[6] The martial gaze is thus inescapably cast on domestic societies increasingly conceived as battlespaces *in potentia* and requiring persistent and intensive surveillance for the purposes of preempting, deterring, and neutralizing emergent threats.[7]

Existing institutional restraints still somewhat constrain the internal deployment of military capabilities, but their loosening can already be felt, not least in the increasing militarization of police forces.[8] On being interrogated in 2013 about whether the president had the power to authorize drone strikes against U.S. citizens on American soil, Attorney General Eric Holder pointedly allowed for the eventuality of "extraordinary circumstance" that would mandate such a use of force.[9] In an era replete with the invocation of "states of exception" to authorize the suspension of prevailing rules and norms, it may not be too fatalistic to presume such an occurrence is merely a matter of time.[10] The last remaining obstacles to the martial gaze's unimpeded global operation are thus slowly but surely being lifted, presaging an age of pervasive lethality.

Tellingly, the "combat cloud" is the new operational concept presently gaining ground in U.S. military circles.[11] The concept is inspired by cloud computing, which pools and distributes computer processing resources and data via the Internet in order to meet the immediate demands of users. Emphasizing the centrality of network connectivity and data management, the combat cloud advances a vision of highly distributed operations that can spontaneously access relevant information and assemble necessary capabilities to nullify any emergent threat, regardless of its location. According to its leading proponent, former Air Force deputy chief of staff for intelligence, surveillance, and reconnaissance David Deptula, the combat cloud's stated objective is to constitute "an omnipresent security complex that is self-forming, and if attacked, self-healing." Such a complex would "enable a deterrent effect that would induce stability wherever employed, or achieve decisive outcomes if force application is actually required."[12] The amorphous and ubiquitous combat cloud may well be the next logical development in armed force most apposite to our "transpolitical logistical epoch."[13]

David Deptula explains that "every object and person . . . is a component, a router, and a node" within the mutable, self-organizing network of the combat cloud, processing and transmitting information at speeds that "enable accurate effects against priority targets."[14] Notwithstanding the accompanying nod to the guiding role of "human understanding,"[15] these designs raise troubling questions regarding the fate of human subjects within global targeting's architecture of control, particularly as available reaction times inexorably contract. Speaking at the annual meeting of the Association of the U.S. Army in 2016, Major General William Hix readily acknowledged the direction in which war is heading: "The speed of events is likely to strain our human abilities. The speed at which machines can make decisions in the far future is likely to challenge our ability to cope, demanding a new relationship between man and machine."[16] The very terms of this relationship will certainly constitute one of the most critical issues of our coming century. Although the future remains necessarily open, we can nevertheless already reflect on the ever-deepening imbrication of humans within the war machine.

The Ghost in the War Machine

In his famous 1938 essay on "The Age of the World Picture," Martin Heidegger sought to identify the metaphysical ground of modern science and technology—that is to say, the specific apprehension of the world that is their very condition of possibility. The "representability" of beings—the presupposition of their accessibility as such to us—was paramount, in the philosopher's view, in realizing the "objectification" of these beings and submitting them to the ambit of calculation and machine technology. From this standpoint, the world appears as a "picture," not merely in the sense that a representational likeness of it is summoned but more fundamentally that any being within the world "stands before us together with what belongs to and stands together with it as a system."[17] For Heidegger, the modern scientific project was therefore synonymous with the imposition of a totalizing, systematic, rational order on the world for the benefit of a human subject conceived as sovereign and domineering. Modernity was accordingly "to be defined by the fact that man becomes the measure and the center of beings. Man is what lies at the bottom of all beings; that is, in modern terms, at the bottom of all objectification and representability."[18]

Rey Chow explicitly builds on Heidegger's analysis with her claim that in the age of aerial bombing, "the world has come to be grasped and conceived as a target—to be destroyed as soon as it can be made visible."[19] In drawing a line between the (visual) objectification of the world and its annihilation, Chow underlines the ubiquitous character of contemporary processes of targeting, whether their lethal force is actualized or remains merely latent, as in the strategy of deterrence. That these processes simultaneously affirm a human subject at their helm is, however, much less clear. It is not merely the case that the positions of subject and object of targeting are inherently interchangeable, contingent on which end of the martial gaze one finds oneself (and admitting that imbalances in the distribution of global power undoubtedly favor certain apportionments of these roles). Indeed, the development of the sociotechnical gaze traced throughout this book also entails the objectification of perception itself as its mechanisms are formalized and externalized. Perception becomes a process without a subject—or at least one that produces its own modes of subjectification. As noted by Hito Steyerl, linear perspective already had the paradoxical effect of simultaneously "empowering the subject by placing it at the center of vision" and undermining "the viewer's individuality by subjecting it to supposedly objective laws of representation."[20] Heidegger himself appears to gesture toward this apparent contradiction when he conceded, soon after World War II, that "even this, the fact that man has turned into the subject and the world the object, is a consequence of the self-establishing essence of technology, not the reverse."[21] Might it then be that the classical sovereign subject was no more than a surface effect of the technologies of perception and representation contemporaneous to it, one that has been progressively shed as these have pursued their development through successive eras of electrification and high-speed computation?

To be sure, this self-determining subject today seems nowhere more precarious than in the military assemblages enmeshing the human ever more deeply within arrays of perceptual prosthetics and dense thickets of cybernetic control loops. In an age of "centaur warfighting" where "hybrid human–machine cognitive architectures will be able to leverage the precision and reliability of automation without sacrificing the robustness and flexibility of human intelligence," identifying the locus of military agency will become ever more difficult.[22] The source of this perilous state of affairs is not to be found in the advent of any particular technology but rather in the unfolding of a general sociotechnical disposition that has been the primary object of our study. This disposition's expression lies first and foremost in its general procedures and diagrams of operation, not in specific material realizations. Thus, today's conduct of targeting can already be seen to sustain a life of its own as a reiterative logistical process. Astrid H. M. Nordin and Dan Öberg observe that the pursuit of a machinic ideal of maximum efficiency and operational perfection dominates contemporary practices of military targeting. Increasingly severed from the logic of means and ends that purports to submit the exercise of armed force to the desiderata of the political, targeting ultimately sees "the subject of warfare dissolve into operationalized repetition."[23]

If our enquiry opened with a camera obscura, it can fittingly close with another martial deployment of the same mechanism. In use from World War I into the 1940s, this particular assemblage assisted the training of aircraft crew in the then novel targeting exercise of aerial bombardment.[24] An instructor ensconced within a dark tent or cabin saw projected on a scale map the light from the sky above through a wide-angle lens embedded in the roof of the structure (Figure 46). As the image of the overflying trainer aircraft moved across the sheet, the operator marked its position at regular intervals provided by the sound

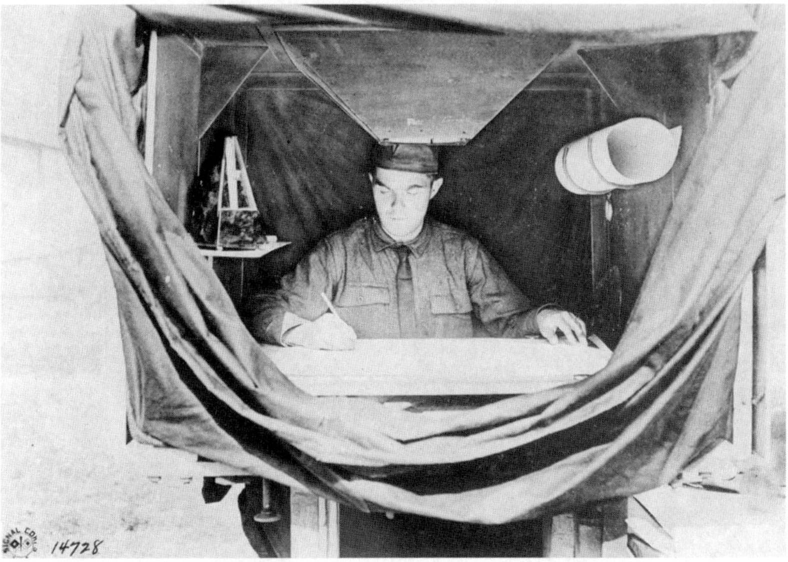

FIGURE 46. *Camera obscura tent for bomber practice, U.S. Air Service bombing school, Clermont-Ferrand, France, 1918. Copyright Imperial War Museums (MH 34216).*

of an adjacent metronome, thereby deducing the aircraft's course and ground speed. The aircraft's mock bomb run on the camera obscura culminated in a visual or radio signal that notified to the instructor the moment of the munition's intended release. The instructor then swiftly calculated the landing position of the virtual bomb and assessed its accuracy.[25] With this information relayed back to the aircraft, every further bomb run strove for an improved aim, honing the operational skills of its crew.

Sitting in complete darkness but for the glow of the illuminated map, diligently registering the movement of an incoming aircraft's image across it to the steady oscillation of a ticking metronome, the camera obscura's occupant is both a passive object of the targeting process and an active if compliant agent tasked with the iterative evaluation and optimization of its performance. Perhaps this duality encapsulates the martial condition we inhabit today, caught between our mobilization within the circulatory networks of the logistics of perception and the roving crosshairs of a global imperium of targeting—and all watched over by machines of glacial indifference.

ACKNOWLEDGMENTS

The travail involved in producing a scholarly book is one mostly endured alone. Yet such an endeavor is also necessarily the outcome of innumerable encounters with other minds and the indispensable task of thinking with, through, and against them in grasping at this world we all share. While some of these minds can be found listed here and there across the text, many more will remain unacknowledged—but not thereby forgotten or unappreciated.

Individual expressions of gratitude must nevertheless go out to Martin Coward, Sam Forsythe, Jairus Grove, Dan Monk, Dan Öberg, John Protevi, Nisha Shah, Benedict Singleton, and Srdjan Vucetic, all of whom read earlier drafts, in part or in whole. I hope I have done some justice to their insightful comments and suggestions.

I would also be remiss not to mention the inspiring students and wonderful colleagues whom I have grown with at Birkbeck throughout the duration of this project's long gestation; their contributions and support have been a ceaseless spur to its completion. Thanks to all at the University of Minnesota Press, but especially Pieter Martin for his faith in the project and the tireless efforts he expended in encouraging me to make the book as good as it could be.

I owe a special debt of loving gratitude to Meerim, whose marital gaze was unfailingly supportive and understanding whenever the book drew me away from precious family time.

Finally, I am endlessly thankful to Mona and Emile for giving me eyes through which to see the world anew.

NOTES

Introduction

1. "The Science of Explosives as Applied to Warlike Purposes," *Nature* 2, no. 29 (May 19, 1870): 50.

2. National Research Council, *U.S. Conventional Prompt Global Strike: Issues for 2008 and Beyond* (Washington, D.C.: National Academies Press, 2008); Amy F. Woolf, "Conventional Prompt Global Strike and Long-Range Ballistic Missiles: Background and Issues," Congressional Research Service, February 24, 2016, https://fas.org/sgp/crs/nuke/R41464.pdf.

3. Robert R. Tomes, *U.S. Defence Strategy from Vietnam to Operation Iraqi Freedom: Military Innovation and the New American War of War, 1973–2003* (London: Routledge, 2007), 67.

4. Martin C. Libicki, "Information and Nuclear RMAs Compared," *Strategic Forum* 82 (Washington, D.C.: Institute for National Strategic Studies, National Defense University Press, 1996), 2.

5. Katharine Hall Kindervater, "The Emergence of Lethal Surveillance: Watching and Killing in the History of Drone Technology," *Security Dialogue* 47, no. 3 (2016).

6. The profusion of scholarship on the drone phenomenon has been such that an exhaustive summation of the field is impossible. Among the most thought-provoking contributions are: Grégoire Chamayou, *A Theory of the Drone,* trans. Janet Lloyd (New York: New Press, 2015); Ian G. R. Shaw, *Predator Empire: Drone Warfare and Full Spectrum Dominance* (Minneapolis: University of Minnesota Press, 2016); Hugh Gusterson, *Drone: Remote Control Warfare* (Cambridge, Mass.: MIT Press, 2016); Kyle Grayson, *Cultural Politics of Targeted Killing: On Drones, Counter-insurgency, and Violence* (London: Routledge, 2017); Derek Gregory, "From a View to a Kill: Drones and Late Modern War," *Theory, Culture and Society* 28, no. 7–8 (2011); Caroline Holmqvist, "Undoing War: War Ontologies and the Materiality of Drone Warfare," *Millennium* 41, no. 3 (2013); Jamie Allison, "The Necropolitics of Drones," *International Political Sociology* 9, no. 2 (2015); Joseph Pugliese, "Drone Casino Mimesis: Telewarfare and Civil Militarization," *Journal of Sociology* 52, no. 3 (2016); Elke Schwarz, "Prescription Drones: On the Techno-Biopolitical Regimes of Contemporary 'Ethical Killing,'" *Security Dialogue* 47, no. 1 (2016); Lauren Wilcox, "Embodying Algorithmic War: Gender, Race, and the Posthuman in Drone Warfare." *Security Dialogue* 48, no. 1 (2017); Katharine Hall Kindervater, "The Technological Rationality of the Drone Strike," *Critical Studies on Security* 5, no. 1 (2017).

7. Derek Gregory, "The Everywhere War," *Geographical Journal* 177, no. 3 (2011).

8. Shaw, *Predator Empire.*

9. Grégoire Chamayou, "The Manhunt Doctrine," *Radical Philosophy* 169 (2011).

10. Noam Lubell and Nathan Derejko, "A Global Battlefield? Drones and the

Geographical Scope of Armed Conflict," *Journal of International Criminal Justice* 11, no. 1 (2013).

11. Frédéric Mégret, "War and the Vanishing Battlefield," *Loyola University Chicago International Law Review* 9, no. 1 (2011).

12. For a related discussion of the changing concept of war in international politics, see Antoine Bousquet, "War," in *Concepts in World Politics,* ed. Felix Berenskoetter (London: SAGE Publications, 2016).

13. Glenn J. Voelz, "The Individualization of American Warfare," *Parameters* 45, no. 1 (2015).

14. Chamayou, *Theory of the Drone,* 54–59.

15. Scott Beauchamp, "The Moral Cost of the Kill Box," *Atlantic,* February 28, 2016, https://www.theatlantic.com/politics/archive/2016/02/the-cost-of-the-kill-box/470751/.

16. *Joint Doctrine for Targeting,* Joint Publication 3-60 (Washington, D.C.: U.S. Joint Chiefs of Staff, January 17, 2002), GL-7.

17. Anna Mulrine, "Warheads on Foreheads," *Air Force Magazine* 91, no. 10 (2008).

18. United States Air Force, "Global Vigilance, Global Reach, Global Power for America" (2014), 8, http://www.airuniversity.af.mil/Portals/10/CMSA/documents/Required_Reading/Global%20Vigilance,%20Reach,%20and%20Power%20for%20America.pdf.

19. Paul Virilio, *Virilio Live: Selected Interviews,* ed. John Armitage (London: SAGE Publications, 2001), 85.

20. John P. Abizaid and Rosa Brooks, *Recommendations and Report of The Task Force on U.S. Drone Policy* (Washington, D.C.: Stimson Center, 2014), 35.

21. Paul Virilio, *War and Cinema: The Logistics of Perception,* trans. Patrick Camiller (1984; London: Verso, 1989).

22. Virilio, *Virilio Live,* 186.

23. The limited recognition of the importance of military logistics is manifest in the relative paucity of scholarly attention given to it, with the two standout analytical engagements being those of Martin van Creveld, *Supplying War: Logistics from Wallenstein to Patton* (Cambridge: Cambridge University Press, 2004), and John A. Lynn, *Feeding Mars: Logistics in Western Warfare from the Middle Ages to the Present* (Boulder, Colo.: Westview Press, 1993). A conceptually rich discussion of logistics in war can also be found in Manuel DeLanda, *War in the Age of Intelligent Machines* (New York: Zone Books, 1991), 105–25.

24. Deborah Cowan, *The Deadly Life of Logistics: Mapping Violence in Global Trade* (Minneapolis: University of Minnesota Press, 2014).

25. Paul Virilio and Sylvère Lotringer, *Pure War* (New York: Semiotext(e), 1983), 103.

26. Antoine-Henri Jomini, *The Art of War,* trans. G. H. Mendell (1838; Mineola, N.Y.: Dover, 2007), 230.

27. "Logistic," Online Etymology Dictionary, http://www.etymonline.com/index.php?term=logistic.

28. For an extensive treatment of the scientific ordering of war, see Antoine Bousquet, *The Scientific Way of Warfare: Order and Chaos on the Battlefields of Modernity* (London: Hurst, 2009). A synthetic overview is provided in Antoine Bousquet, "Chaoplexic Warfare or the Future of Military Organization," *International Affairs* 84, no. 5 (2008).

29. Nigel Thrift, "Movement-Space: The Changing Domain of Thinking Resulting from the Development of New Kinds of Spatial Awareness," *Economy and Society* 33, no. 4 (2004): 589.

30. Jomini, *Art of War,* 231.

31. Within this study, the meaning given to the term of "targeting" differs from that associated with its official usage in the military, where it denotes the decisionary process by which targets are selected and prioritized, and weapon systems allocated to them, all in accordance with overall command objectives and prevailing rules of engagement. Instead, "targeting" will refer here to the various calculative operations and sociotechnical concatenations that, through the conjoining of organs of perception and weaponry, make such deliveries of force possible in the first place. *Department of Defense Dictionary of Military and Associated Terms,* Joint Publication 1-02 (November 8, 2010, amended February 15, 2016), 236, https://fas.org/irp/doddir/dod/jp1_02.pdf.

32. "Do you not see that the eye embraces the beauty of the world? The eye is commander of astronomy; it makes cosmography; it guides and rectifies all the human arts; it conducts man to various regions of the world; it is the prince of mathematics; its sciences are most certain; it has measured the height and size of the stars; it has disclosed the elements and their distributions; it has made predictions of future events by means of the course of the stars; it has generated architecture, perspective, and divine painting. Oh excellent above all other things created by God . . . And it triumphs over nature, in that the constituent parts of nature are finite, but the works that the eye commands of the hands are infinite, as is demonstrated by the painter in his rendering of numberless forms of animals, grasses, trees, and places." Martin Kemp, *Leonardo* (Oxford: Oxford University Press, 2011), 51.

33. David Michael Levin, ed., *Modernity and the Hegemony of Vision* (Berkeley: University of California Press, 1993), 212.

34. Martin Jay, *Downcast Eyes: The Denigration of Vision in Twentieth-Century French Thought* (Berkeley: University of California Press, 1994), 69.

35. Antonio Somaini, "On the Scopic Regime," *Leitmotiv* 5 (2005–6): 36.

36. Martin Jay has been keen to emphasize the existence of alternative scopic regimes present in modern "visual subcultures," among which the expressions of seventeenth-century Dutch art and the baroque that respectively broke with the precepts of strict perspectivalism. As important as these distinct cultural forms may have been, they cannot be said to have significantly affected the development of a militarized vision, where the hegemony of the original perspectival regime endured into the twentieth century. The advent of electromagnetic theory and its associated sensorial technologies has since displaced the centrality of geometrical optics but without thereby abandoning the project of mastery through rationalized perception and correlated spatialization. Jay, "Scopic Regimes of Modernity," in *Vision and Visuality,* ed. Hal Foster (Seattle, Wash.: Bay Press, 1988).

37. Gregory, "From a View to a Kill," 193.

38. Grayson, *Cultural Politics of Targeted Killing,* 153.

39. Martin Coward, "Networks, Nodes and De-territorialised Battlespace: The Scopic Regime of Rapid Dominance," in *From Above: War, Violence and Verticality,* ed. Peter Adey, Mark Whitehead, and Alison J. Williams (London: Hurst, 2013), 99.

40. Henry George Liddell and Robert Scott, *A Greek–English Lexicon* (Oxford: Clarendon Press, 1940).

41. Jacques Lacan, *The Seminar of Jacques Lacan: The Four Fundamental Concepts of Psychoanalysis (Book XI),* ed. Jacques-Alain Miller, trans. Alan Sheridan (1973; London: Hogarth Press, 1977); Michel Foucault, *The Birth of the Clinic: An Archaeology of Medical Perception,* trans. Alan Sheridan (1963; New York: Pantheon Books, 1973); Laura Mulvey, "Visual Pleasure and Narrative Cinema," *Screen* 16, no. 3 (1975).

42. Hermann von Helmholtz, *Popular Lectures on Scientific Subjects,* trans. E. Atkinson (London: Longmans, Green, 1873), 202.

43. Von Helmholtz, *Popular Lectures,* 219.

44. Paul Virilio, *The Virilio Reader,* ed. James Der Derian (Oxford: Blackwell, 1998), 8–9.

45. John Pettegrew, *Light It Up: The Marine Eye for Battle in the War for Iraq* (Baltimore, Md.: Johns Hopkins University Press, 2015), 8.

46. Lucy Suchman, "Situational Awareness: Deadly Bio-convergence at the Boundaries of Bodies and Machines," *MediaTropes* 5, no. 1 (2015).

47. Virilio, *Virilio Live,* 43.

48. Donald MacKenzie, *Inventing Accuracy: A Historical Sociology of Nuclear Missile Guidance* (Cambridge, Mass.: MIT Press, 1990).

49. Martin Shaw, *The New Western Way of War: Risk-Transfer War and Its Crisis in Iraq* (Cambridge: Polity Press, 2005).

50. Of course, for Virilio, the "accident" is an integral feature of any technology that is not only ineradicable but also a functional constituent of the development of technique. Paul Virilio, *The Original Accident,* trans. Julie Rose (Cambridge: Polity Press, 2007).

51. Eyal Weizman, *The Least of All Possible Evils: Humanitarian Violence from Arendt to Gaza* (London: Verso, 2011); Patricia Owens, "Accidents Don't Just Happen: The Liberal Politics of High-Technology 'Humanitarian' War," *Millennium* 32, no. 3 (2003).

52. Maja Zehfuss, "Targeting: Precision and the Production of Ethics," *European Journal of International Relations* 17, no. 3 (2011); Chamayou, *Theory of the Drone,* 140–49.

53. John Nagl, *Learning to Eat Soup with a Knife: Counterinsurgency Lessons from Malaya and Vietnam* (Chicago: University of Chicago Press, 2005); David Kilcullen, *Counterinsurgency* (London: Hurst, 2010).

54. H. Bruce Franklin, *War Stars: The Superweapon and the American Imagination* (Amherst: University of Massachusetts Press, 2008).

55. The CEP for any given weapons system does of course entail that half of the strikes conducted will fall outside the stated radius, the spread of their deviation dependent on the system's probability distribution. Derek Gregory, "War and Peace," *Transactions of the Institute of British Geographers* 35, no. 2 (2010): 173.

56. Chad Harris, "The Omniscient Eye: Satellite Imagery, 'Battlespace Awareness,' and the Structures of the Imperial Gaze," *Surveillance and Society* 4, no. 1/2 (2006): 118.

57. For a more detailed discussion of the conceptualization of technology and society informing this book, see Antoine Bousquet, "A Revolution in Military Affairs? Changing Technologies and Changing Practices of Warfare," in *Technology and World Politics: An Introduction,* ed. Daniel R. McCarthy (New York: Routledge, 2017).

58. Bernard Stiegler, *The Decadence of Industrial Democracies: Disbelief and Discredit,* trans. Daniel Ross and Suzanne Arnold (Cambridge: Polity Press, 2011), 1:51.

59. Lorenzo C. Simpson, *Technology, Time, and the Conversations of Modernity* (London: Routledge, 1995), 17.

60. Gilles Deleuze and Félix Guattari, *A Thousand Plateaus: Capitalism and Schizophrenia,* trans. Brian Massumi (Minneapolis: University of Minnesota Press, 1987). See also Michele Acuto and Simon Curtis, eds., *Reassembling International Theory: Assemblage Thinking and International Relations* (London: Palgrave, 2014).

61. Bruno Latour, *Pandora's Hope: Essays on the Reality of Science Studies* (Cambridge, Mass.: Harvard University Press, 1999), 214.

62. To take just the military context, see Mike Bourne, "Guns Don't Kill People,

Cyborgs Do: A Latourian Provocation for Transformatory Arms Control and Disarmament," *Global Change, Peace and Security* 24, no. 1 (2012); Jairus Grove, "An Insurgency of Things: Foray into the World of Improvised Explosive Devices," *International Political Sociology* 10, no. 4 (2016); Benjamin Meiches, "Weapons, Desire and the Making of War," *Critical Studies on Security* 5, no. 1 (2017); Nisha Shah, "Gunning for War: Infantry Rifles and the Calibration of Lethal Force," *Critical Studies on Security* 5, no. 1 (2017).

63. Gilbert Simondon, *On the Mode of Existence of Technical Objects,* trans. Cecile Malaspina and John Rogove (Minneapolis: University of Minnesota Press, 2016).

64. Mark Salter, ed., *Making Things International 1* (Minneapolis: University of Minnesota Press, 2015); Mark Salter, ed., *Making Things International 2* (Minneapolis: University of Minnesota Press, 2016).

65. Andrew Pickering, *The Mangle of Practice: Time, Agency, and Science* (Chicago: University of Chicago Press, 1995), 19.

1. Perspective

1. William M. Ivins, *On the Rationalization of Sight* (New York: Da Capo Press, 1973), 13.

2. Ivins, *On the Rationalization of Sight,* 9.

3. Bruno Latour, "Visualization and Cognition: Thinking with Eyes and Hands," *Knowledge and Society* 6 (1986): 10.

4. Edwin Hutchins, "Material Anchors for Conceptual Blends," *Journal of Pragmatics* 37, no. 10 (2005).

5. Ivins, *On the Rationalization of Sight,* 9–10.

6. Christopher S. Wood, introduction to *Perspective as Symbolic Form,* by Erwin Panofsky, trans. Christopher S. Wood (New York: Zone Books, 1991), 13.

7. Ken Hillis, "The Power of Disembodied Imagination: Perspective's Role in Cartography," *Cartographica* 31, no. 3 (1994): 7.

8. William M. Ivins, *Art and Geometry: A Study in Space Intuitions* (New York: Dover, 1946), 32.

9. Perspective images can have more than one vanishing point toward which sets of parallels converge—up to three for the representation of a three-dimensional Euclidean space from a fixed point within it, although constructions with a greater number of vanishing points are also possible. Crucially, the same fundamental principles of projective geometry pertain to all cases, and single-point perspective can therefore rightly be considered as archetypal.

10. James Burke, *The Day the Universe Changed* (London: Little, Brown, 1995), 76.

11. Ivins, *Art and Geometry,* 13.

12. Kim H. Veltman, "Ptolemy and the Origins of Linear Perspective," in *Atti del Convegno Internazionale di Studi: La Prospettiva Rinascimentale, Milan, 1977,* ed. Marisa Dalai-Emiliani (Florence: Centro Di, 1980), 403–7. See also Veltman's extensive bibliographical review on *The Sources of Perspective* (2004) at http://sumscorp.com/img/file/2004_Sources_of_Perspective.pdf.

13. Panofsky, *Perspective as Symbolic Form,* 43.

14. Richard Tarnas, *The Passion of the Western Mind* (London: Pimlico, 1991), 230. See also Margaret Wertheim, *The Pearly Gates of Cyberspace: A History of Space from Dante to the Internet* (New York: Norton, 1999).

15. Samuel Edgerton, *The Renaissance Rediscovery of Linear Perspective* (New York: Harper & Row, 1976), 7.

16. Brunelleschi reportedly used polished silver for the sky in the painting, thereby reflecting the drifting skies alongside the image of the Baptistery.

17. Edgerton, *Renaissance Rediscovery*; Martin Kemp, "Science, Non-science and Nonsense: The Interpretation of Brunelleschi's Perspective," *Art History* 1, no. 2 (1978); Shigeru Tsuji, "Brunelleschi and the Camera Obscura: The Discovery of Pictorial Perspective," *Art History* 13, no. 3 (1990).

18. Martin Kemp, *The Science of Art: Optical Themes in Western Art from Brunelleschi to Seurat* (New Haven, Conn.: Yale University Press, 1990), 167–68.

19. "The distinction between them is one of degree. In surveying it is often merely a question of finding two points in order to establish the height of a tower or breadth of a river etc. In a perspective picture we need to fix many more points in order to arrive at a three-dimensional outline of the object(s) in front of us." Kim Veltman, "Military Surveying and Topography: The Practical Dimension of Renaissance Linear Perspective," *Revista da Universidade de Coimbra* 27 (1979): 330.

20. Bas C. van Fraassen, *Scientific Representation: Paradoxes of Perspective* (Oxford: Oxford University Press, 2010), 60.

21. Kemp, *Science of Art,* 167.

22. Alberti, quoted in Patricia Lee Rubin, *Images and Identity in Fifteenth-Century Florence* (New Haven, Conn.: Yale University Press, 2007), 93.

23. Joel Snyder, "Picturing Vision," *Critical Inquiry* 6, no. 3 (1980): 515. Alberti's text reveals a robust knowledge of optical theory, then known as *perspectiva naturalis* and based on the classical works of Euclid and Ptolemy and subsequent contributions of tenth-century Arab scholar Alhazen. Alberti is, however, careful to sidestep long-standing controversies within optics (such as whether visual rays enter the eye or are issued from it) to focus entirely on the geometric character of vision and its practical applications for the systematic representation of space.

24. Leon Battista Alberti, *On Painting,* trans. John R. Spencer (New Haven, Conn.: Yale University Press, 1966), 90.

25. Alberti, *On Painting,* 46–47.

26. Paolo Freguglia, "Reflections on the Relationship between Perspective and Geometry in the Sixteenth and Seventeenth Centuries," *Nexus Network Journal* 11, no. 3 (2009): 333.

27. The Latin root of the term "perspective," *perspicere,* is particularly apposite here, with its meaning of "seeing through." Alberti, *On Painting,* 56.

28. Alberti, *On Painting,* 57.

29. Hubert Damisch, *The Origin of Perspective,* trans. John Goodman (Cambridge, Mass.: MIT Press, 1994), 332.

30. David L. Martin, *Curious Visions of Modernity: Enchantment, Magic, and the Sacred* (Cambridge, Mass.: MIT Press, 2011), 145.

31. John Berger, *Ways of Seeing* (London: Penguin Books, 1972), 16. See also Hannah B. Higgins's assertion that "this empowered viewer embodies a new kind of human subjectivity, one that would make possible the ideal of the modern genius, an individual capable of changing the course of history, science, or culture." Higgins, *The Grid Book* (Cambridge, Mass.: MIT Press, 2009), 149.

32. Burke, *The Day the Universe Changed,* 77.

33. One can see in the perspectival moment the presaging of the epistemological torsion described by Michel Foucault in his account of the posterior emergence of new knowledges in which "man appears in his ambiguous position as an object of knowledge and as a subject that knows: enslaved sovereign, observed spectator." Foucault, *The Order of Things: Archaeology of the Human Sciences* (London: Routledge, 2005), 340.

34. Friedrich Kittler, *Optical Media,* trans. Anthony Enns (Cambridge: Polity Press, 2010), 19–20.

35. William J. Mitchell, *The Reconfigured Eye: Visual Truth in the Post-photographic Era* (Cambridge, Mass.: MIT Press, 1992), 117–18.

36. Lev Manovich, "The Automation of Sight: From Photography to Computer Vision," in *Electronic Culture: Technology and Visual Representation,* ed. Timothy Druckery (New York: Aperture Foundation, 1996), 239.

37. Edgerton sees in the use of a grid structure evidence of the influence on the development of linear perspective of the recently rediscovered *Geographia* from Ptolemy, particularly its use of longitudinal and latitudinal lines to allow for the proportionate representation of geographic space. "The grid-formed *velo* was not merely a device for transferring a scale drawing. It was a means for organising the visible world itself into a geometric composition, structured on evenly spaced grid coordinates." Edgerton further argues that the third method of projection proposed by Ptolemy for flattening the globe onto a two-dimensional surface constitutes an early formulation of the geometric principles underpinning linear perspective. Edgerton, *Renaissance Rediscovery,* 119.

38. Van Fraassen, *Scientific Representation,* 65.

39. Erwin Panofsky, "Dürer as a Mathematician," in *The World of Mathematics,* ed. James R. Newman (London: Novello, 1960), 1:610.

40. Marshall McLuhan, *Counterblast* (London: Rapp & Whiting, 1970), 11.

41. Giacomo Barozzi da Vignola, *Le Due Regole della Prospettiva Practica* (1583), cited in Veltman, "Military Surveying and Topography," 333–34.

42. Veltman, "Military Surveying and Topography," 338.

43. Filippo Camerota, "Renaissance Descriptive Geometry: The Codification of Drawing Methods," in *Picturing Machines, 1400–1700,* ed. Wolfgang Lefèvre (Cambridge, Mass.: MIT Press, 2004), 182.

44. Joan Gadol, *Leon Battista Alberti: Universal Man of the Early Renaissance* (Chicago: University of Chicago Press, 1963), 71.

45. Massimo Scolari, *Oblique Drawing: A History of Anti-perspective,* trans. Jenny Condie Palandri (Cambridge, Mass.: MIT Press, 2012).

46. Judith V. Field and Jeremy Gray, eds., *The Geometrical Work of Girard Desargues* (New York: Springer-Verlag, 1987).

47. Panofsky, *Perspective as Symbolic Form,* 70. Gaspard Monge termed his own approach "descriptive geometry," which he developed in solving design problems related to fortifications and taught from 1795 in the newly formed French military engineering school, the École Polytechnique. In the program of his lectures, he outlined the basic purpose of descriptive geometry: "First it allows one to represent three-dimensional objects susceptible of being rigorously defined on a two-dimensional drawing. . . . Second . . . by taking the description of such objects to its logical conclusion, we can deduce something about their shape and relative positioning." Monge, *Landmark Writings in Western Mathematics, 1640–1940,* ed. Ivor Grattan-Guinness and Roger Cooke (Amsterdam: Elsevier, 2005), 226.

48. Johann H. Lambert, *Die freye Perspektive oder Anweisung jeden perspektivischen Aufriß von freyen Stücken und ohne Grundriß zu verfertigen* (Zurich, 1759).

49. Abraham Bosse, *Le Peintre Converti aux Précises et Universelles Règles de son Art* (Paris, 1637), quoted in Damisch, *Origin of Perspective,* 150.

50. Lev Manovich, "Modern Surveillance Machines: Perspective, Radar, 3-D Computer Graphics and Computer Vision," in *CTRL (Space): Rhetorics of Surveillance from Bentham to Big Brother,* ed. T. Y. Levin (Cambridge, Mass.: MIT Press, 2002), 383.

2. Sensing

1. Among the other sensory organs called upon in close-quarters combat is the inner ear's vestibular system, which is crucial to the body's sense of balance and spatial orientation.

2. Albert van Helden, "The Invention of the Telescope," *Transactions of the American Philosophical Society* 67, no. 4 (1977); Albert van Helden, Sven Dupré, Rob van Gent, and Huib Zuidervaart, eds., *The Origins of the Telescope* (Amsterdam: KNAW Press, 2010).

3. Rolf Willach, "The Long Route to the Invention of the Telescope," *Transactions of the American Philosophical Society* 98, no. 5 (2008).

4. Roger Bacon, *The Opus Majus of Roger Bacon* (1267; Philadelphia: University of Pennsylvania Press, 1928), 582.

5. Eileen Reeves, *Galileo's Glassworks: The Telescope and the Mirror* (Cambridge, Mass.: Harvard University Press, 2008), 27. Another much-rehearsed tale speaks of a mirror located within the great lighthouse of Alexandria that doubled as a powerful emitter of light and telescopic sight so that, as one twelfth-century chronicler put it, when "ships came to the city from Greece or the West with warlike aims, [they] could be seen when they were still twenty days' travel away, such that the inhabitants were able to mount a defence" (23–24).

6. Maurice of Nassau was notably responsible for the systematic introduction of drill, one of the key features in the mechanistic transformation of warfare in the early modern era. See Bousquet, *Scientific Way of Warfare,* 53–62.

7. Huib J. Zuidervaart, "The 'True Inventor' of the Telescope: A Survey of 400 Years of Debate," in van Helden et al., *Origins of the Telescope,* 14.

8. Engel Sluiter, "The Telescope before Galileo," *Journal for the History of Astronomy* 28, no. 3 (1997): 226.

9. Albert van Helden, introduction to *Siderus Nuncius, or The Sideral Messenger,* by Galileo Galilei (1610; Chicago: University of Chicago Press, 1989), 7–8.

10. Reeves, *Galileo's Glassworks,* 11.

11. Richard Dunn, *The Telescope: A Short History* (London: National Maritime Museum, 2009), 32–33.

12. Martin van Creveld, *Command in War* (Cambridge, Mass.: Harvard University Press, 1985), 10–11.

13. Martin Pegler, *Sniper Rifles: From the 19th to the 21st Century* (Oxford: Osprey, 2010), 8.

14. R. Mullineux Walmsley, "Presidential Address to the Optical Society (22 March 1900)," *Transactions of the Optical Society* 1 (1899): 35–36.

15. Arthur Conan Doyle, *The Great Boer War* (1900; London: Floating Press, 2011), 254.

16. Contracts between the preeminent British manufacturer Barr & Stroud and the admiralty amounted to an expenditure of £313,000 between 1907 and 1913, compared to a mere £47,000 with the War Office over the same period. Stephen Curtis Sambrook, "The Optical Munitions Industry in Great Britain, 1888–1923" (Ph.D. diss., University of Glasgow, 2005), 103.

17. Roy and Kay MacLeod, "War and Economic Development: Government and the Optical Industry in Britain, 1914–18," in *War and Economic Development,* ed. J. M. Winter (Cambridge: Cambridge University Press, 1975), 171.

18. Sambrook, *Optical Munitions Industry,* 186. Toward the end of the nineteenth century, Galilean designs for binoculars increasingly gave way to prismatic models

that offered a wider field of view and greater magnification—between 8× and 16× the naked eye by the outbreak of World War I compared to the former's modest 2× or 4×. As for the prismatic field telescopes of the day, the most powerful attained as much as 32× magnification in ideal conditions.

19. "Les Instruments d'Optique Employés aux Armées," *Bulletin de la Société d'Encouragement pour l'Industrie Nationale* (1919): 611. For more on the French optics industry during World War I, see Charlotte Bigg, "L'Optique de Précision et la Première Guerre Mondiale," *Schweizerische Zeitschrift für Geschichte* 55 (2005).

20. H. C. McKay, "Magic Eyes of War and Peace," *Smith's Weekly,* November 18, 1939, 13.

21. Virilio, *War and Cinema,* 26.

22. Virilio, *War and Cinema,* 3.

23. Martin Pegler, *The Military Sniper since 1914* (Oxford: Osprey, 2001), 14. For a colorful firsthand recounting of British experiences with sniping and scouting during World War I by one of its most prominent promoters, see H. Hesketh-Prichard, *Sniping in France—with Notes on the Scientific Training of Scouts, Observers, and Snipers* (London: Hutchinson, 1920).

24. Between 2007 and 2011, the U.S. Army commissioned almost 200,000 units of the 4× advanced combat optical gunsight for fitting on the M4 carbine and M16 assault rifle.

25. At the greatest ranges, wind speed and direction become significant factors in addition to bullet drop.

26. Geoffrey Parker, *The Military Revolution: Military Innovation and the Rise of the West, 1500–1800* (Cambridge: Cambridge University Press, 1988).

27. Although larger devices provide greater accuracy over longer distances by increasing the base length of the range-finding triangle, their size naturally restricts their mobility and renders them vulnerable to enemy spotting, thus limiting their deployment to the navy and rear positions. The largest optical range finder ever built was the Barr & Stroud FZ assembled for the coastal defense of Portsmouth in 1923 with a 100-foot (30 meter) base and could reliably measure distances to objects as distant as 30 kilometers. Michael Moss and Iain Russell, *Range and Vision: The First Hundred Years of Barr & Stroud* (Edinburgh: Mainstream, 1988), 112.

28. Stereoscopic range finders were generally found to be more accurate for mobile and indistinct targets than their coincidence counterparts, but the demands made upon the operator's visual system rendered their skillful use more challenging. In any case, either design represented a marked improvement on the earlier optical range finders that were deployed in the last decades of the nineteenth century, such as the British Army's Watkin Mekometer, which required two observers wielding a box sextant and optical square connected to each other by a taut cord.

29. "Nature's range-finder is found in the binocular vision with which we are endowed, the pupillary distance being the base, and the muscular accommodation the means by which the range is unconsciously estimated. For short distances by long practice we have more or less acquired an accurate perception of range, hence it is that in shooting game in the field the practised shot will estimate both direction and range automatically. For the longer distances employed in war, this Nature's range-finder breaks down, more especially as the opportunities for training it are not sufficiently numerous, and also because for long distances varying atmospheric conditions so enormously alter the visual effect produced on the retina. Optical range-finders are therefore a necessity in modern warfare, and hundreds of inventors have applied their talents to the solution of the problem." Walmsley, "Presidential Address," 86–87.

30. These include air temperature and density, wind, rotation of the earth, parallax and differences in elevation between the positions of the guns and the sighting systems, and the internal and external ballistics of specific guns and projectiles.

31. David A. Mindell, *Between Human and Machine: Feedback, Control, and Computing before Cybernetics* (Baltimore, Md.: Johns Hopkins University Press, 2003), 26–27.

32. Mindell, *Between Human and Machine,* 20–21.

33. "[The] basic principle [of the mechanical range keeper] is that if present range is determined, and if the rate and direction at which range is changing can be computed from the speeds and courses of target and own ship, then the range at a later instant can be determined by computing the change of range for any convenient time interval and adding this change of range to the original value of observed present range, assuming that the range rate remains constant during the period of the computation. This computation can be used to keep the range up to date while the [fire-control] problem is being solved, and to predict for change during the time of flight of the projectile." *Naval Ordnance and Gunnery, Volume 2: Fire Control* (U.S. Navy, Bureau of Personnel, NavPers 10798, 1958), chap. 15.

34. Stuart Bennett, "A Brief History of Automatic Control," *IEEE Control Systems Magazine* 16, no. 3 (1996): 20.

35. Mindell, *Between Human and Machine,* 221.

36. Placing great hopes in the Norden bombsight's accuracy, the U.S. military purchased up to 90,000 units and trained bomber crews extensively in their use. In the end, however, the bombsight never performed in battlefield conditions at the promised levels of accuracy; policy eventually reverted to area bombing. While the general principle of operation behind the device was never called into question, the sight's operational shortcomings point to the recurrent difficulties encountered by new targeting technologies in replicating on the battlefield the performance achieved under controlled conditions.

37. Zabet Patterson, "From the Gun Controller to the Mandala: The Cybernetic Cinema of John and James Whitney," *Grey Room* 36 (2009): 42.

38. Patterson, "From the Gun Controller to the Mandala," 42.

39. Richard Langton Gregory, *The Eye and The Brain: The Psychology of Seeing* (New York: McGraw-Hill, 1966), 93–94.

40. Jordan Crandall, "Precision + Guided + Seeing," CTheory.net, January 1, 2006, http://www.ctheory.net/articles.aspx?id=502.

41. In his work under the auspices of the OSRD, mathematician and future founder of cybernetics Norbert Wiener devised a statistical technique that explicitly recognized these sources of uncertainty and proposed a means to handle them. Although the significant improvements in predictive accuracy it yielded were still insufficient to decisively resolve the antiaircraft problem, Wiener's groundbreaking method nonetheless subsequently found widespread applications, both military and civilian, in the tracking of stochastic phenomena, notably in the areas of vehicle navigation and guidance. Wiener, "Statistical Method of Prediction in Fire Control" (Washington, D.C.: Office of Scientific Research and Development, National Defense Research Committee, Division 7, December 1942), http://www.dtic.mil/dtic/tr/fulltext/u2/a800106.pdf.

42. National Defense Research Committee, *Guided Missiles and Techniques,* Summary Technical Report of Division 5 (Washington, D.C.: National Defense Research Committee, 1946), 198–99.

43. B. F. Skinner, "Pigeons in a Pelican," *American Psychologist* 15, no. 1 (1960): 30.

44. Skinner further describes a later improvement on the three-pigeon system in

which "the majority took on a more characteristically democratic function" by ensuring that the dissenting bird was "punished for his minority opinion", thereby reinforcing the desired behavior. Skinner, "Pigeon in a Pelican," 31.

45. Skinner, "Pigeons in a Pelican," 36.

46. James H. Capshew, "Engineering Behavior: Project Pigeon, World War II, and the Conditioning of B. F. Skinner," *Technology and Culture* 34, no. 4 (1993): 851.

47. Alexander B. Magoun, *Television: The Life Story of a Technology* (Westport, Conn.: Greenwood Press, 2007), 79.

48. V. K. Zworykin, "Flying Torpedo with an Electric Eye," *RCA Review* 7 (1946): 294.

49. Albert Abramson, *The History of Television, 1942–2000* (Jefferson, N.C.: McFarland, 2008), 9.

50. For an overview of American efforts at television-guided weaponry during World War II, see Katherine Fehr Chandler, "Drone Flight and Failure: The United States' Secret Trials, Experiments, and Operations in Unmanning, 1936–1973" (Ph.D. diss., University of California, Berkeley, 2014), 34–57.

51. National Defense Research Committee, *Guided Missiles and Techniques,* 4.

52. Norbert Wiener, *The Human Use of Human Beings: Cybernetics and Society* (London: Free Association Books, 1989), 23.

53. Of lesser but growing currency are ultraviolet sensors that perform particularly well in snowy terrain, given icy water's high reflection of ultraviolet light. Ultraviolet sensors also complement infrared heat-seeking weapon systems, such as the portable Stinger surface-to-air missile, by recording the mixture of smoke particles from the exhaust of a tracked vehicle and thereby distinguishing it from any potential flare decoys.

54. Adam K. Raymond, "'We Own the Night': The Rise and Fall of the US Military's Night-Vision Dominance," *Task and Purpose,* July 11, 2017, https://taskandpurpose.com/night-rise-fall-us-militarys-night-vision-dominance/.

55. The common choice of the color green for night-vision displays is dictated by the human eye's greater sensitivity to variations in shade at these frequencies of the visible spectrum.

56. Pegler, *Military Sniper since 1914,* 47.

57. W. N. Arnquist, "Survey of Early Infrared Developments," *Proceedings of the Institute of Radio Engineers* 47, no. 9 (1959).

58. James A. Ratches, Richard Chait, and John W. Lyons, "Some Recent Sensor-Related Army Critical Technology Events," Center for Technology and National Security Policy, National Defense University (February 2013), 14.

59. Avi Parush, Michelle S. Gauthier, Lise Arseneau, and Denis Tang, "The Human Factors of Night Vision Goggles: Perceptual, Cognitive, and Physical Factors," *Reviews of Human Factors and Ergonomics* 7, no. 1 (2011).

60. "Night-Vision Contact Lenses May Be in Your Future," *Army Times,* March 28, 2014, http://archive.armytimes.com/article/20140328/NEWS04/303280045/Night-vision-contact-lenses-may-your-future.

61. R. Bruce Lindsay, "The Story of Acoustics," *Acoustical Society of America* 39, no. 4 (1966).

62. Alongside aerial reconnaissance, the other main technique for locating artillery during World War I was flash spotting: observers used powerful optical lenses to measure the bearing of the muzzle flashes of enemy guns relative to several positions. Once plotted, the intersection of at least three distinct bearings indicated the location of the hostile battery.

63. Roy MacLeod, "Sight and Sound on the Western Front: Surveyors, Scientists,

and the 'Battlefield Laboratory,' 1915–1918," *War and Society* 18, no. 1 (2000): 43–45. For an extensive treatment of French and Italian sound-ranging efforts, see Martina Schiavon, *Itinéraires de la precision: geodesiens, artilleurs, savants et fabricants d'instruments de precision en France, 1870–1930* (Nancy: Presses Universitaires de Nancy, 2013), 425–599.

64. Michael V. Namorato, "A Concise History of Acoustics in Warfare," *Applied Acoustics* 59, no. 2 (2000): 109.

65. J. S. Finan and W. J. Hurley, "McNaughton and Canadian Operational Research at Vimy," *Journal of the Operational Research Society* 48, no. 1 (1997): 13.

66. Peter Chasseaud, "Field Survey in the Salient: Cartography and Artillery Survey in the Flanders Operations in 1917," in *Passchendaele in Perspective: The Third Battle of Ypres,* ed. Peter Liddle (London: Leo Cooper, 1997), 120. Although the scales of these respective wartime projects are of a different order, the British efforts were certainly able to draw on some of the brightest scientific minds available, including Nobel Prize laureate Lawrence Bragg and several members of Ernest Rutherford's research team from Manchester University. For their own part, the French drew on the talents of prominent mathematician and future prime minister Paul Painlevé. On the German side, Max Born, later known for his pioneering work in quantum physics, participated in their endeavors.

67. John Shiga, "Sonar: Empire, Media, and the Politics of Underwater Sound," *Canadian Journal of Communication* 38, no. 3 (2013): 366–67.

68. Gary E. Weir, "The American Sound Surveillance System: Using the Ocean to Hunt Soviet Submarines, 1950–1961," *International Journal of Naval History* 5, no. 2 (2006): 1.

69. Shiga speaks of the training of active sonar technicians in terms of the creation of "new listening subjects" tasked with an aural "hermeneutics of echoes," citing a World War II Royal Navy instructor's insistence that operators should "keep in mind what a submarine echo may sound like under all conditions and so distinguish it from other almost exactly similar sounds." Shiga, "Sonar," 369–71.

70. Weir, "American Sound Surveillance System," 8.

71. BBN Technologies's Boomerang shooter-detection system uses an array of seven microphones distributed around a single mast to localize fire at ranges of up to 150 meters within a second of a shot being detected. The system was notably deployed in Iraq and Afghanistan by American and British forces.

72. Techniques for passive radar that do not rely on emitting pulsed signals but instead process the reflections of signals originating from other sources of radio transmission (such as broadcasting and telecommunications radio masts) are currently being developed. While military-grade systems are still in a trial phase, their eventual successful deployment would make the use of radar tracking undetectable and potentially negate a whole raft of existing stealth technologies. Arend G. Westra, "Radar versus Stealth: Passive Radar and the Future of U.S. Military Power," *Joint Force Quarterly* 55, 4th Quarter (2009).

73. David A. Mindell, "Automation's Finest Hour: Radar and System Integration in World War II," in *Systems, Experts, and Computers: The Systems Approach in Management and Engineering, World War II and After,* ed. Agatha C. Hughes and Thomas P. Hughes (Cambridge, Mass.: MIT Press, 2000).

74. Yves Blanchard, *Le Radar, 1904–2004: Histoire d'un Siècle d'Innovations Techniques et Opérationnelles* (Paris: Ellipses, 2004), 282.

75. Robert Buderi, *The Invention That Changed the World: How a Small Group of*

Radar Pioneers Won the Second World War and Launched a Technological Revolution (New York: Simon & Schuster, 1996), 249.

76. The use of a unique pulse coding for individual target designator laser beams serves to ward against deliberate jamming or misdirection; it also permits the simultaneous designation of multiple targets to which specific guided munitions can be allocated.

77. Max Boot, "From Saigon to Desert Storm: How the U.S. Military Reinvented Itself after Vietnam," *American Heritage* 57, no. 6 (November/December 2006), http://www.americanheritage.com/content/saigon-desert-storm.

78. Lee Hutchinson, "Bullseye from 1,000 Yards: Shooting the $17,000 Linux-Powered Rifle," Ars Technica, April 1, 2013, http://arstechnica.com/gadgets/2013/03/bullseye-from-1000-yards-shooting-the-17000-linux-powered-rifle/.

79. Brian Anderson, "Long Shot: Inside the Scope of Smart Weapons," Motherboard, September 23, 2013, http://motherboard.vice.com/blog/long-shot-inside-the-scope-of-smart-weapons.

80. "Sandia's Self-Guided Bullet Prototype Can Hit Target a Mile Away," Sandia Labs News Release, January 30, 2012, https://share.sandia.gov/news/resources/news_releases/bullet/#.ViYRqyuZVMm.

81. Evan Ackerman, "DARPA's Self-Steering EXACTO Bullets Home In on Moving Targets," *IEEE Spectrum,* April 29, 2015, http://spectrum.ieee.org/tech-talk/aerospace/military/darpas-selfsteering-bullets-home-in-on-moving-targets.

82. William J. Fanning, "The Historical Death Ray and Science Fiction in the 1920s and 1930s," *Science Fiction Studies* 37, no. 2 (2010).

83. Robert W. Seidel, "From Glow to Flow: A History of Military Laser Research and Development," *Historical Studies in the Physical and Biological Sciences* 18, no. 1 (1987): 114.

84. U.S. Department of State, *Blinding Laser Weapons (Protocol IV),* weapons protocol between the United States and other governments, adopted October 13, 1995, https://www.state.gov/documents/organization/190580.pdf.

85. In fact, the risk of irreversible ocular damage is considerably increased if laser radiation passes through magnifying optics such as binoculars before striking a human eye.

86. Bengt Anderberg and Myron L. Wolbarsht, *Laser Weapons: The Dawn of a New Military Age* (New York: Plenum Press, 1992), 94–98.

87. The emitting device is, however, likely to generate considerable heat and accordingly possess a distinctive infrared signature.

88. Columba Peoples, *Justifying Ballistic Missile Defence: Technology, Security and Culture* (Cambridge: Cambridge University Press, 2010), 119.

89. Seidel, "From Glow to Flow," 145.

90. Peter Goodchild, *Edward Teller: The Real Dr. Strangelove* (Cambridge, Mass.: Harvard University Press, 2004), 365.

91. A further potential application of the airborne laser was as an antisatellite weapon that could damage or destroy low-orbit reconnaissance satellites.

92. Katie Drummond, "RIP, Raygun: Pentagon's Laser Plane Laid to Rest," *Wired,* February 21, 2012, http://www.wired.com/2012/02/airborn-laser-rip/.

93. Sydney J. Freedberg, "Return of the ABL? Missile Defense Agency Works on Laser Drone," *Breaking Defense,* August 17, 2015, http://breakingdefense.com/2015/08/return-of-the-abl-missile-defense-agency-works-on-laser-drone/.

94. Recent U.S. programs include the Army's THEL (Tactical High-Energy Laser)

and ZEUS laser systems, the Air Force's ATL (Advanced Tactical Laser), the Navy's LaWS (Laser Weapon System), and DARPA's HELLADS (High-Energy Liquid Laser Area Defense System).

3. Imaging

1. Jay, *Downcast Eyes,* 65.
2. Kittler, *Optical Media,* 42.
3. McLuhan, *Counterblast,* 11.
4. Mitchell, *Reconfigured Eye,* 28.
5. Snyder, "Picturing Vision," 511.
6. Snyder, "Picturing Vision," 512.
7. Kittler, *Optical Media,* 63.
8. Mitchell, *Reconfigured Eye,* 27.
9. Lorraine Daston and Peter Galison, "The Image of Objectivity," in "Seeing Science," special issue, *Representations,* no. 40 (Fall 1992).
10. Jean Massin, ed., *Zola Photographe* (Paris: Hoëbeke, 1990), 11.
11. William Henry Fox Talbot, "The Pencil of Nature" (1844), in *Henry Fox Talbot: Selected Texts and Bibliography,* ed. Mike Weaver (Boston: G. K. Hall, 1993), 92.
12. Walter Benjamin, "Little History of Photography," trans. Edmund Jephcott and Kingsley Shorter, in *Walter Benjamin: Selected Writings, Volume 2: 1927–1934,* ed. Michael W. Jennings, Howard Eiland, and Gary Smith (Cambridge, Mass.: Harvard University Press, 1999): 512.
13. Jens Andermann, *The Optic of the State: Visuality and Power in Argentina and Brazil* (Pittsburgh, Pa.: University of Pittsburgh Press, 2007), 199.
14. The cinematic image is technically nothing more than the serialization of photography, restoring motion from the same device that arrests it by the sequential display of successive still images at an interval beneath the human visual system's threshold of perceptibility.
15. Ernst Jünger, *On Pain,* trans. David C. Durst (1934; New York: Telos Press, 2008), 39. For treatments of Jünger's wider conception of the role of technology in modern war, see Antoine Bousquet, "Ernst Jünger and the Problem of Nihilism in the Age of Total War," *Thesis Eleven* 132, no. 1 (2016), and Bogdan Costea and Kostas Amiridis, "Ernst Jünger, Total Mobilisation and the Work of War," *Organization* 24, no. 4 (2017).
16. Jünger, *On Pain,* 39.
17. Jünger, *On Pain,* 38–40.
18. Jünger, *On Pain,* 40.
19. François Arago, *Rapport sur le Daguerréotype* (Paris: Bachelier, 1839), 25–31 (my translation).
20. Edward Said, *Orientalism* (New York: Penguin Classics, 2003), 127. See also Ali Behdad and Luke Gartlan, eds., *Photography's Orientalism: New Essays on Colonial Representation* (Los Angeles, Calif.: Paul Getty Trust, 2013).
21. Manovich, "Automation of Sight," 231.
22. Oliver Wendell Holmes, "The Stereoscope and the Stereograph," *Atlantic Monthly* 3, no. 20 (1859), https://www.theatlantic.com/magazine/archive/1859/06/the-stereoscope-and-the-stereograph/303361/.
23. Étienne-Jules Marey, "Le Fusil Photographique," *La Nature* 464 (April 22, 1882): 327.
24. Virilio, *War and Cinema,* 15.
25. Friedrich Kittler, *Gramophone, Film, Typewriter,* trans. Geoffrey Winthrop-Young and Michael Wutz (Stanford, Calif.: Stanford University Press, 1999), 124.

26. Ernst Jünger, "War and Photography," *New German Critique* 59 (1993): 24.
27. Harun Farocki, *Imprint Writings* (New York: Lukas & Sternberg, 2001), 188.
28. Arago, *Rapport sur le Daguerréotype,* 31.
29. Colonel Laussedat, *Instruments et Appareils Iconométriques et Métrophotographiques des Collections des Arts et Métiers* (Paris: Imprimerie Nationale, 1893), 19.
30. Laussedat, *Instruments et Appareils,* 1.
31. Laussedat, *Instruments et Appareils,* 2–3.
32. As one contemporary American promoter of the technique put it, "If we delegate to the photographic camera the duty of making a permanent record of nature, as seen from two or more points of view, the intersection of the rays of light, reaching those points of view respectively from the same objects, as pictured in photographs properly placed, will, by their intersection on paper to become a map, give the positions of those objects relatively, as plotted, to the points of view and to one another." R. Meade Bache, "Civil and Military Photogrammetry," *Proceedings of the American Philosophical Society* 30, no. 138 (1892): 230–31.
33. John Adolphus Flemer, *An Elementary Treatise on Phototopographic Methods and Instruments, Including a Concise Review of Executed Phototopographic Surveys and of Publications on this Subject* (New York: Wiley, 1906), 12–13.
34. Jan von Brevern, "Fototopografia: The 'Futures Past' of Surveying," *Intermediality: History and Theory of the Arts, Literature and Technologies* 17 (2011).
35. From 1885, Meydenbauer served as the head of the newly founded Royal Prussian Photogrammetric Institute and was tasked with photographic surveying of the Prussian state's cultural monuments. By 1920, the institute had made 20,000 photogrammetric images of around 2,600 sites.
36. One of the most notable terrain surveys was Édouard-Gaston Deville's extensive mapping of the Canadian Rocky Mountains in the late ninteenth century using a refined phototopographic technique derived from Laussedat's pioneering work. See E. Deville, *Photographic Surveying, Including the Elements of Descriptive Geometry and Perspective* (Ottawa: Government Printing Bureau, 1895).
37. Peter Chasseaud, *Artillery's Astrologers: A History of British Survey and Mapping on the Western Front, 1914–1918* (Lewes, U.K.: Mapbooks, 1999), 4.
38. Denis Cosgrove and William L. Fox, *Photography and Flight* (London: Reaktion Books, 2010), 26.
39. Beaumont Newhall, *Airborne Camera: The World from the Air and Outer Space* (London: Focal Press, 1969), 46.
40. Walter Benjamin, "The Work of Art in the Age of Its Technological Reproducibility: Second Version" (1936), trans. Edmund Jephcott and Harry Zohn., in *Walter Benjamin: Selected Writings, Volume 3: 1935–1938,* ed. Howard Eiland and Michael W. Jennings (Cambridge, Mass.: Harvard University Press, 2002), 132.
41. Some of the grounds for early reticence at the employment of aerial photography, such as those recalled in the memoirs of a Royal Flying Corps officer, appear particularly quaint today: "Aerial photography was a thing that had not been done fore, and to expose the whole set-up behind the enemy's lines was to invade a privacy that had always been accorded the enemy, never mind that it happened because nobody could do anything about it." Terrence J. Finnegan, *Shooting the Front: Allied Air Reconnaissance in the First World War* (Stroud, U.K.: History Press, 2011), 51.
42. Charles J. Cohen, "Early History of Remote Sensing," in *Proceedings of the 29th Applied Imagery Pattern Recognition Workshop* (Washington, D.C.: IEEE Computer Society, 2000), 7.
43. Finnegan, *Shooting the Front,* 114.
44. Finnegan, *Shooting the Front,* 7.

45. Live aerial observation did continue to have a use for the real-time reporting of artillery fire and troop maneuvers through wireless radio communication with the ground.

46. Finnegan, *Shooting the Front,* 251.

47. Peter Adey, *Aerial Life: Spaces, Mobilities, Affects* (Oxford: Wiley-Blackwell, 2010), 89.

48. "Aerial Photography and the War," *Scientific Monthly* 6, no. 4 (1918): 384.

49. Sinai, Palestine, and Mesopotamia were notably the sites of particularly intensive topographic mapping from aerial photographs during the war. See Peter Collier, "Innovative Military Mapping Using Aerial Photography in the First World War: Sinai, Palestine and Mesopotamia, 1914–1919," *Cartographic Journal* 31, no. 2 (1994).

50. The underlying principles for the geometric correction (or orthorectification) of aerial images through the application of one of Girard Desargues's theorems on projective geometry had been demonstrated by Austrian naval officer Theodor Scheimpflug before the war, but the techniques available during the conflict could only yield approximate results.

51. One postwar account would go so far as to proclaim the stereoscope the "deadliest weapon of the war." Douglas Reid, "The Eye behind the Lines," *Popular Mechanics* 31, no. 4 (April 1919): 529.

52. Jonathan Crary sees in the stereoscope one of the paradigmatic technologies of a new understanding of subjective, embodied vision in the nineteenth century, marking an epistemic rupture with the disincarnate perspectivalism associated with Renaissance painting and the monocular camera obscura. The rise of modern bourgeois subjectivity and its associated cultures of consumerism and spectatorship are thus held to have accompanied the establishment of physiological optics. Crary's assertion that "stereoscopic relief or depth has no unifying logic or order" and that "if perspective implied a homogeneous and potentially metric space, the stereoscope discloses a fundamental disunified and aggregate field of disjunct elements" may well hold at the level of the observer's phenomenal experience of these images (125). But it occludes the fact that stereoscopy was concurrently assimilated into a general regime of rationalized vision and geometrical optics, supplementing rather than undercutting the visual metricization of space advanced by the perspectival image. The photogrammetric use of stereoscopic images to extract measurements of physical distances and the coexistence of monocular and stereoscopic models of range finders illustrate this complementarity well enough in the military context. The scientific and cultural developments traced by Crary remain nonetheless highly significant because they laid essential ground for the deeper integration of the human sensorium within perceptual assemblages in the twentieth century—or as Cary puts it himself, "Efficiency and rationalization in many areas of human activity depended on information about the capacities of the human eye." Crary, *Techniques of the Observer: On Vision and Modernity in the Nineteenth Century* (Cambridge, Mass.: MIT Press, 1990), 16.

53. Paul K. Saint-Amour, "Modernist Reconnaissance," *Modernism/Modernity* 10, no. 2 (2003): 356.

54. Saint-Amour, "Modernist Reconnaissance," 365.

55. Peter Collier, "The Impact on Topographic Mapping of Developments in Land and Air Survey: 1900–1939," *Cartography and Geographic Information Science* 29, no. 3 (2002).

56. Cosgrove and Fox, *Photography and Flight,* 55.

57. Roy M. Stanley, *World War II Photo Intelligence* (London: Sidgwick & Jackson, 1981), 3.

58. Stanley, *World War II Photo Intelligence,* 315.

59. Dwayne A. Day, John M. Logsdon, and Brian Latell, eds., *Eye in the Sky: The Story of the Corona Spy Satellites* (Washington, D.C.: Smithsonian Books, 1998), 1.

60. As Philip Windsor points out, arms control efforts during the Cold War only strove to rationalize the possession and configuration of armaments so as to perpetuate the stabilization of deterrence. Arms control was therefore an integral part of the system of deterrence and was not in any sense synonymous with actual disarmament. Windsor, *Strategic Thinking: An Introduction and Farewell* (Boulder, Colo.: Lynne Rienner, 2002).

61. Curtis Peebles, *The Corona Project: America's First Spy Satellites* (Annapolis, Md.: Naval Institute Press, 1997), 170–71.

62. Robert A. McDonald, "Corona: A Success for Space Reconnaissance, a Look into the Cold War, and a Revolution for Intelligence," *Photogrammetric Engineering and Remote Sensing* 51, no. 6 (1995): 719.

63. Peter A. Gorin, "Zenit: The Soviet Response to Corona," in Day, Logsdon, and Latell, *Eye in the Sky,* 157–70.

64. J. E. Lipp, R. M. Salter Jr., R. S. Wehner, R. R. Carhart, and C. R. Culp, *Utility of a Satellite Vehicle for Reconnaissance* (Santa Monica, Calif.: RAND Corporation, 1951), 12.

65. The National Reconnaissance Office was created in 1961 to oversee the design and operation of U.S. spy satellites, but its very existence was not officially acknowledged until 1992.

66. Cosgrove and Fox, *Photography and Flight,* 73.

67. The United Kingdom possesses no independent military satellite reconnaissance capability, receiving its imagery primarily through its intelligence agreements with the United States.

68. Warren Ferster, "U.S. Government Eases Restrictions on DigitalGlobe," Space News, June 11, 2014, http://spacenews.com/40874us-government-eases-restrictions-on-digitalglobe/.

69. A typical reconnaissance satellite in a low-earth circular orbit of 500 kilometers travels at around 7.5 kilometers per second, completing its orbit roughly every ninety minutes.

70. U.S. Army Space and Missile Defense Command/Army Forces Strategic Command, "Kestrel Eye Fact Sheet" (2012), https://www.smdc.army.mil/FactSheets/KESTRELEYE.pdf; Jason B. Cutshaw, "Army Deploys Kestrel Eye Satellite," Army.mil, October 25, 2017, https://www.army.mil/article/195548/army_deploys_kestrel_eye_satellite.

71. An added benefit of higher orbits is a reduced vulnerability to antisatellite weaponry.

72. "DARPA's MOIRE: Video Scud Hunts from Space," *Defense Industry Daily,* September 6, 2011, http://www.defenseindustrydaily.com/DARPAs-MOIRE-Video-Scud-Hunts-from-Space-07079/.

73. Tim Blackmore, *War X: Human Extensions in Battlespace* (Toronto: University of Toronto Press, 2005), 139–40.

74. For a considered discussion of UAVs and their place within the wider scopic regimes of contemporary war, see Gregory, "From a View to a Kill."

75. Noel Schachtman, "Air Force to Unleash 'Gorgon Stare' on Squirting Insurgents," *Wired,* February 19, 2009, http://www.wired.com/dangerroom/2009/02/gorgon-stare/.

76. BAE Systems, ARGUS-IS brochure, http://www.baesystems.com/cs/groups/public/documents/document/mdaw/mtg2/~edisp/baes_162769.pdf. The ideal platform

for ARGUS-IS may well turn out to be the long-endurance, solar-powered UAVs currently in development and which promise a flight autonomy of months, if not years.

77. Paul Virilio, *La Machine de Vision* (Paris: Editions Galilée, 1988), 126.

78. Harun Farocki, "Phantom Images," in "Localities," ed. Saara Liinamaa, Janine Marchessault, and Christine Shaw, special issue, *Public* 29 (2004): 17. See also Farocki's series of films *Eye/Machine I* (2001), *Eye/Machine II* (2002), and *Eye/Machine III* (2003).

79. Von Brevern, "Fototopografia," 64.

80. Trevor Paglen, "Invisible Images (Your Pictures Are Looking at You)," New Inquiry, December 8, 2016, https://thenewinquiry.com/invisible-images-your-pictures-are-looking-at-you/.

81. Mitchell, *Reconfigured Eye.*

82. Edmond Couchot, "Image Puissance Image," *Revue d'Esthétique* 7 (1984): 124.

83. John J. May, "Sensing: Preliminary Notes on the Emergence of Statistical-Mechanical Geographic Vision," in *Perspecta 40 "Monster": The Yale Architectural Journal,* ed. Marc Guberman, Jacob Reidel, and Frida Rosenberg (Cambridge, Mass.: MIT Press, 2008), 46.

84. Photogrammetry has also been assisted in recent decades by the advent of in-flight GPS permitting the determination of sensor positions to unprecedented levels of accuracy.

85. While such boundaries are eminently porous, a distinction can be made between photogrammetry and computer vision, according to which the former is primarily concerned with the production of maps and strives for the most precise possible recording of spatial dimensions through imaging, while the latter tends to focus on object recognition and change detection within a scene, with a correspondingly higher tolerance of measurement imprecision.

86. Hubert L. Dreyfus, *What Computers Still Can't Do: A Critique of Artificial Reason* (Cambridge, Mass.: MIT Press, 1992); John Haugeland, *Artificial Intelligence: The Very Idea* (Cambridge, Mass.: MIT Press, 1989); Jack Copeland, *Artificial Intelligence: A Philosophical Introduction* (Oxford: Blackwell, 1993); Reza Negarestani, *Intelligence and Spirit* (Falmouth, U.K.: Urbanomic, 2018).

87. DeLanda, *War in the Age of Intelligent Machines,* 200–203.

88. Jon R. Carr and James S. Sobek, "Digital Scene Matching Area Correlator (DSMAC)," *Image Processing for Missile Guidance: Proceedings of the Society of Photo-Optical Instrumentation Engineers* 238 (1980); Geoffrey B. Irani and James P. Christ, "Image Processing for Tomahawk Scene Matching," *Johns Hopkins APL Technical Digest* 15, no. 3 (1994).

89. The U.S. Navy presently holds a stockpile of around 3,500 Tomahawk missiles at an estimated combined worth of $2.6 billion, with the most sophisticated models coming at a unit cost of $1.5 million.

90. "A pixel value is a sample in time and space of light intensities projected onto the picture plane." Mitchell, *Reconfigured Eye,* 66.

91. Commonly used image features include edges, corners, lines, curves, and other geometrical shapes, the choice of which is largely determined by the specific task. In addition to these features, computer vision systems also make use of other cues such as motion, shading, and texture to infer the three-dimensional characteristics of a given scene. See Emmanuelle Truco and Alessandro Verri, *Introductory Techniques for 3-D Computer Vision* (Upper Saddle River, N.J.: Prentice Hall, 1998).

92. D. L. Milgram and C. M. Bjorklund, "3-D Model Matching For Missile Guidance," *Image Processing for Missile Guidance: Proceedings of the Society of Photo-Optical Instrumentation Engineers* 238 (1980).

93. Lawrence G. Roberts, "Machine Perception of Three-Dimensional Solids" (Ph.D. thesis, Massachusetts Institute of Technology, 1963), https://dspace.mit.edu/handle/1721.1/11589.

94. Spice, an Israeli bomb guidance system that converts unguided bombs into precision-guided munitions, can be preloaded with images and geolocational coordinates of up to a hundred targets. Once a specific target has been allocated and the warhead has been released, the system applies scene-matching algorithms in order to pair the reference data to the live input it receives from its electro-optical or infrared sensor. Having established a positive match, the weapon proceeds to home in onto its target autonomously within an advertized circular error probability of 3 meters. Rafael Advanced Defense Systems, Spice brochure, https://web.archive.org/web/20160303204716/http://www.rafael.co.il/marketing/SIP_STORAGE/FILES/4/924.pdf.

95. "Mind's Eye," DARPA, Information Innovation Office, 2012, http://www.homelandsecuritynewswire.com/dr20121030-smart-camera-to-describe-what-it-sees-and-reason-about-what-it-cannot-see.

96. U.N. General Assembly, "Report of the Special Rapporteur on Extrajudicial, Summary or Arbitrary Executions, Christof Heyns," April 9, 2013, A/HRC/23/47, http://www.ohchr.org/Documents/HRBodies/HRCouncil/RegularSession/Session23/A-HRC-23-47_en.pdf.

97. James A. Ratches, "Review of Current Aided/Automatic Target Acquisition Technology for Military Target Acquisition Tasks," *Optical Engineering* 50, no. 7 (2011).

98. Israel's Iron Dome or America's THAAD are examples of existing air defense systems with autonomous targeting capabilities.

99. John Johnston, "Machinic Vision," *Critical Inquiry* 26, no. 1 (1999): 45.

100. Johnston, "Machinic Vision," 29.

101. Dora Dougherty Strother and Hubert W. Upton, "Head-Mounted Display/Control System in V/STOL Operations," paper presented at the 27th Annual National V/STOL Forum of the American Helicopter Society, Washington, D.C., May 1971, 1, http://archives.rotor.com/documents/PUBLICATIONS/DOC102209-033.pdf.

102. Heads-up displays grew out of the gyroscopic reflector gunsights deployed on aircraft during World War II. The sights assisted the targeting of enemy planes through the illuminated display of a moving reticle that automatically factored in the necessary variables of angle of deflection, bullet drop, and range for an accurate aim. John Kim submits that the British Mark II gunsight can be thought of as "the first see-through graphical interface, because it was the first 'computing' device to employ a projection screen on which graphical information was displayed that enhanced operators' performance at their task." Kim, "The Origin of the See-Through Graphical Interface: World War II Aircraft Gunsights and the Status of the Material in Early Computer Interface Design," *Convergence* 21, no. 2 (2015): 219.

103. Michel Friedman, Kathrin Friedrich, Moritz Queisner, and Christian Stein, "Conceptualizing Screen Practices: How Head-Mounted Displays Transform Action and Perception," *MediaTropes* 6, no. 1 (2016).

104. Joseph A. Birt and Thomas A. Furness, "Visually Coupled Systems," *Air University Review* 20, no. 3 (March–April 1974).

105. Gerald F. Marshall, "Back from the Past: The Helmet Integrated System of Albert Bacon Pratt (1916)," *Helmet-Mounted Displays: Proceedings of the Society of Photo-Optical Instrumentation Engineers* 1116 (1989): 3.

106. U.S. Patent 1,183,492 awarded to Albert B. Pratt (May 16, 1916), 5.

107. Dictated primarily by prevailing engineering constraints and the necessity

of minimizing total helmet weight, the IHADSS's monocular design means that display information is only delivered to the user's right eye. The consequent requirement for users to juggle between visual inputs demands intensive training, and original concerns over the strain it places on pilots' visual systems have persisted until today. Biocular arrangements (in which both eyes receive a single identical image) or binocular designs (where each eye is presented with a slightly different perspective of the same image, allowing for stereoscopic depth perception) can mitigate or resolve many of these problems but are invariably more costly and technically challenging to implement. Clarence E. Rash, "A 25-Year Retrospective Review of Visual Complaints and Illusions Associated with a Monocular Helmet-Mounted Display," *Displays* 29, no. 2 (2008).

108. Adam Ciralsky, "Will It Fly?," *Vanity Fair,* September 16, 2013, http://www.vanityfair.com/politics/2013/09/joint-strike-fighter-lockheed-martin.

109. Lev Manovich, *The Language of New Media* (Cambridge, Mass.: MIT Press, 2001), 97. A literal disappearance of the screen occurs with virtual retinal display technologies that use lasers or light-emitting diodes to project images directly onto the retina. John R. Lewis, "In the Eye of the Beholder," *IEEE Spectrum,* May 3, 2004, http://spectrum.ieee.org/biomedical/imaging/in-the-eye-of-the-beholder.

110. Ivan E. Sutherland, "Sketchpad: A Man–Machine Graphical Communication System" (Ph.D. diss., Massachusetts Institute of Technology, 1963).

111. Ivan E. Sutherland, "A Head-Mounted Three Dimensional Display," in *Proceedings of the AFIPS Fall Joint Computer Conference* (Washington, D.C.: Thompson Books, 1968), 757. The algorithmic automation of perspectival imaging necessary for the production of three-dimensional computer graphics merely applies the inverse procedure of that which permits the machine recognition and identification of three-dimensional objects. Thus, as Manovich underlines, "The two fields were born simultaneously: 3-D computer graphics and computer vision, automation of imaging and sight." Manovich, "Automation of Sight," 233.

112. Dean F. Kocian, "A Visually-Coupled Airborne Systems Simulator (VCASS)—An Approach to Visual Simulation" (Wright-Patterson Air Force Base, Ohio: Aerospace Medical Research Laboratory, 1977), http://oai.dtic.mil/oai/oai?verb=getRecord&metadataPrefix=html&identifier=ADA039999.

113. Kocian, "Visually-Coupled Airborne Systems Simulator," 3.

114. Thomas A. Furness, "The Super Cockpit and Human Factors Challenges," *Proceedings of the Human Factors and Ergonomics Society Annual Meeting* 30, no. 1 (1986): 48.

115. Thomas A. Furness and Dean F. Kocian, "Putting Humans into Virtual Space" (Seattle: University of Washington, Human Interface Technology Laboratory, 1986). Just as some of the earlier flight simulators used physical terrain models and video cameras, the synthetic landscapes presented to Super Cockpit pilots were three-dimensional computer models generated from a "stored-terrain data base" of digitized Department of Defense maps. Howard Rheingold, *Virtual Reality* (London: Martin Secker & Warburg, 1991), 206.

116. Furness, "Super Cockpit," 48.

117. Woodrow Barfield, Craig Rosenberg, and Thomas A. Furness, "Situation Awareness as a Function of Frame of Reference, Computer-Graphics Eyepoint Elevation, and Geometric Field of View," *International Journal of Aviation Psychology* 5, no. 3 (1995).

118. Furness and Kocian, "Putting Humans into Virtual Space."

119. One UAV application is that of picture-in-picture, in which the standard video imagery is displayed with a surrounding border consisting of the corresponding

synthetically generated environment, thereby increasing the operator's field of view. Gloria L. Calhoun, Mark H. Draper, Mike F. Abernathy, Frank Delgado, and Michael Patzek, "Synthetic Vision System for Improving Unmanned Aerial Vehicle Operator Situation Awareness," in "Enhanced and Synthetic Vision," ed. Jacques G. Verly, special issue, *Proceedings of SPIE,* no. 5802 (2005), https://pdfs.semanticscholar.org/3d3e/c04351101476d2fb619d55b034101f87290d.pdf.

120. Ivan E. Sutherland, "The Ultimate Display," in *Multimedia: From Wagner to Virtual Reality,* ed. Randall Parker and Ken Jordan (New York: Norton, 2001), 236.

4. Mapping

1. Anders Engberg-Pedersen, *Empire of Chance: The Napoleonic Wars and the Disorder of Things* (Cambridge, Mass.: Harvard University Press, 2015), 145.

2. David Harvey, *Spaces of Capital: Towards a Critical Geography* (New York: Routledge, 2001), 220.

3. Nigel Thrift, *Non-representational Theory: Space, Politics, Affect* (London: Routledge, 2008), 94–95.

4. Jeremy Black, *The Power of Knowledge: How Information and Technology Made the Modern World* (New Haven, Conn.: Yale University Press, 2014), 73.

5. John B. Harley, "Deconstructing the Map," *Cartographica* 26, no. 2 (1989); Denis Wood, *The Power of Maps* (New York: Guilford Press, 1992); John Pickles, *A History of Spaces: Cartographic Reason, Mapping and the Geo-coded World* (London: Routledge, 2006); Jeremy Crampton, *Mapping: A Critical Introduction to Cartography and GIS* (Chichester, U.K.: Wiley-Blackwell, 2010).

6. Of course, the projection of a three-dimensional sphere onto a two-dimensional surface inevitably entails some visual distortion (of shape, area, and/or directionality), and the Gall-Peters map proposed to replace the Mercator map is no exception. Peters's claim to have produced the only projection ever to be needed cannot therefore be sustained. His underlying objection to the ubiquity of the Mercator projection and the Eurocentric bias that his own map was intended to redress does, however, retain greater salience. See Jeremy W. Crampton, "Cartography's Defining Moment: The Peters Projection Controversy, 1974–1990," *Cartographica* 31, no. 4 (1994).

7. Jeremy W. Crampton, "Maps as Social Constructions: Power, Communication and Visualization," *Progress in Human Geography* 25, no. 2 (2001): 240.

8. Yves Lacoste, *La géographie, ça sert, d'abord, à faire la guerre* (Paris: La Découverte, 2012), 57.

9. The title Mercator gave to his 1569 map described it as "ad usum navigatium" (for use in navigation).

10. Hutchins, "Material Anchors for Conceptual Blends."

11. The notion of "user" is here understood in the widest possible sense and is not restricted to designating human agents. As this chapter details, maps (or at least the geospatial data that compose them) can also very well support the tasks of nonhuman agents, such as self-guided cruise missiles.

12. David Buisseret, *The Mapmakers' Quest: Depicting New Worlds in Renaissance Europe* (Oxford: Oxford University Press, 2003), 115.

13. Denis Wood, *Rethinking the Power of Maps* (New York: Guilford Press, 2010), 28.

14. Martin van Creveld, *Technology and War* (New York: Free Press, 1989), 116–18.

15. David Buisseret, ed., *Monarchs, Ministers, and Maps: The Emergence of Cartography as a Tool of Government in Early Modern Europe* (Chicago: University of Chicago Press, 1992).

16. Jean Lefort, *L'aventure cartographique* (Paris: Éditions Belin, 2004), 153. This

kind of cartographic dissimulation persists to this day, with national governments ensuring that publicly available sources such as Google Maps do not display sites deemed critical to national security. It is almost certainly within the former Soviet Union that the art of cartographic deception reached its apex, going so far as to use for its civilian maps a special cartographic projection that introduced random variations that made impossible any recovery of the true geography of the places represented. Greg Miller, "Inside the Secret World of Russia's Cold War Mapmakers," *Wired,* July 2015, http://www.wired.com/2015/07/secret-cold-war-maps/.

17. W. A. Seymour, ed., *A History of the Ordnance Survey* (Folkestone, U.K.: Wm. Dawson & Sons, 1980); Rachel Hewitt, *Map of a Nation: A Biography of the Ordnance Survey* (London: Granta, 2010).

18. G. R. Crone, *Maps and Their Makers* (London: Hutchinson House, 1953), 155.

19. For an overview of the growing production and usages of military maps in this period, see Jeremy Black, "A Revolution in Military Cartography? Europe, 1650–1815," *Journal of Military History* 73, no. 1 (2009).

20. Although aspiring to the rigor and systematicity of triangulated surveys, maps produced hastily in the midst of military operations in unfamiliar territory were frequently much more approximate affairs. As one French topographer and general put it in 1803, under these conditions, "procedures become approximations, drawings are transformed into sketches, and the descriptions are no more than reconnaissances." For all this, such rough surveys were still able to fulfill the practical purpose of facilitating the movement and placement of troops. Anders Engberg-Pedersen, "Sketching War: August von Larisch's Collection of Field Maps from the Russian Campaign of 1812," *Imago Mundi* 66, no. 1 (2014): 72.

21. Alexander Kott, Douglas J. Peters, and Stephen Riese, "A Journey into the Mind of Command: How DARPA and the Army Experimented with Command in Future Warefare," in *Battle of Cognition: The Future Information-Rich Warfare and the Mind of the Commander,* ed. Alexander Kott (Westport, Conn.: Praeger Security International, 2008), 62.

22. Engberg-Pedersen, *Empire of Chance,* 157.

23. Philipp von Hilgers, *War Games: A History of War on Paper,* trans. Ross Benjamin (Cambridge, Mass.: MIT Press, 2012).

24. Tomas J. Bassett, "Cartography and Empire Building in Nineteenth-Century West Africa," *Geographical Review* 84, no. 3 (1994); Matthew H. Edney, *Mapping an Empire: The Geographical Construction of British India, 1765–1843* (Chicago: University of Chicago Press, 1997); Thongchai Winichakul, *Siam Mapped: A History of the Geo-Body of a Nation* (Honolulu: University of Hawai'i Press, 1994).

25. Jeremy Black, *Maps and Politics* (London: Reaktion Books, 1997), 150.

26. Henry Edgar Eames, *Military Maps Explained* (Kansas City, Mo.: Franklin Hudson, 1908), 9.

27. Herbert M. Wilson, *Topographic Surveying, Including Geographic, Exploratory, and Military Mapping* (New York: Wiley, 1905), 92–107.

28. Van Creveld, *Command in War,* 289n128.

29. Patrick O'Sullivan and Jesse W. Miller, *The Geography of Warfare* (Beckenham, U.K.: Croom Helm, 1983), 21.

30. MacLeod, "Sight and Sound on the Western Front," 24.

31. Lefort, *L'aventure cartographique,* 199.

32. For a detailed review of British survey efforts, see Chasseaud, *Artillery's Astrologers.* On French endeavors, see Schiavon, *Itinéraires de la precision,* 347–420.

33. Chasseaud, *Artillery's Astrologers,* 499.

34. Chasseaud, *Artillery's Astrologers*, 16.

35. The need for advancing troops to follow close behind an artillery barrage so as to gain the greatest tactical advantage from the disruption of the enemy's defensive positions exposed them to considerable risks of fratricide. The French are said to have held that if 10 to 15 percent of their casualties were not caused by their own barrage, then the troops were not near enough to it. Jonathan B. A. Bailey, *Field Artillery and Firepower* (Annapolis, Md.: Naval Institute Press, 2004), 251.

36. Lt. Col. E. M. Jack cited in Derek Gregory, "Gabriel's Map: Cartography and Corpography in Modern War," in *Geographies of Knowledge and Power*, ed. Peter Meusburger, Derek Gregory, and Laura Suarsana (Dordrecht, Netherlands: Springer, 2015), 117.

37. Peter Chasseaud, "British, French and German Mapping and Survey on the Western Front in the First World War," in *Fields of Battle: Terrain in Military History*, ed. Peter Doyle and Matthew R. Bennett (Dordrecht, Netherlands: Kluwer Academic, 2002), 172.

38. William Rankin, *After the Map: Cartography, Navigation, and the Transformation of Territory in the Twentieth Century* (Chicago: University of Chicago Press, 2016), 125.

39. Collier, "Impact on Topographic Mapping."

40. Chasseaud, *Artillery's Astrologers*, 16.

41. Rankin, *After the Map*, 171.

42. Peter Chasseaud, *Mapping the Second World War* (Glasgow: Collins, 2015), 10.

43. Black, *Power of Knowledge*, 323.

44. Crone, *Maps and Their Makers*, 154.

45. Rankin, *After the Map*, 85.

46. Joseph A. Russell, "Military Geography," in *American Geography: Inventory and Prospect*, ed. Preston E. James and Clarence F. Jones (Syracuse, N.Y.: Syracuse University Press, 1954), 487.

47. An exception is made for the polar regions that use the Universal Polar Stereographic (UPS) coordinate system because of the inadequacy of the Mercator projection for the representation of these areas of the globe.

48. A geodetic datum is defined as "a set of constants specifying the coordinate system used for geodetic control, i.e., for calculating the coordinates of points on the Earth." *Geodetic Glossary* (Rockville, Md.: National Geodetic Survey, National Ocean Service, National Oceanic and Atmospheric Administration, 1986), 54.

49. John Cloud, "Imaging the World in a Barrel: Corona and the Clandestine Convergence of the Earth Sciences," *Social Studies of Science* 31, no. 1 (2001): 244.

50. *Geodesy for the Layman* (Philadelphia, Pa.: Defense Mapping Agency, 1984), 5.

51. Deborah Jean Warner, "Political Geodesy: The Army, the Air Force, and the World Geodetic System of 1960," *Annals of Science* 59, no. 4 (2002): 374.

52. According to the Defense Mapping Agency's own technical manual, "The World Geodetic System is a geocentric system that provides a basic reference frame and geometric figure for the earth, models the earth gravimetrically, and provides the means for relating positions on various datums to an earth-centered, earth fixed coordinate system." *Defense Mapping Agency Technical Manual 8358.1: Datums, Ellipsoids, Grids, and Grid Reference Systems* (Philadelphia, Pa.: Defense Mapping Agency, 1990), http://earth-info.nga.mil/GandG/publications/tm8358.1/tr83581a.html.

53. The Soviet Union devised its own global geodetic reference system underpinned by the same basic design principles. Its latest version, PZ-90, currently supports the GLONASS (Global Navigation Satellite System) satellite navigation system.

54. The deployment of Russia's GLONASS system was disrupted by the collapse of the Soviet Union, but its subsequent revival saw it attain full global capacity on par

with GPS in 2011. The European Union and China are also working on their own respective global navigation satellite systems, and a number of other countries, such as India and Japan, are pursuing regional systems. For a detailed account of the complex entanglement and synergetic development of GPS and geodesy, see Alan G. Evans, Robert W. Hill, Geoffrey Blewitt, Everett R. Swift, Thomas P. Yunck, Ron Hatch, Stephen M. Lichten, Stephen Malys, John Bossler, and James P. Cunningham, "The Global Positioning System Geodesy Odyssey," *Navigation* 49, no. 1 (2002).

55. Laura Kurgan, *Close Up at a Distance: Mapping, Technology and Politics* (Brooklyn, N.Y.: Zone Books, 2013), 69.

56. Stuart Elden, "Secure the Volume: Vertical Geopolitics and the Depth of Power," *Political Geography* 34, no. 1 (2013).

57. Robert Watson-Watt cited in William Rankin, "The Geography of Radionavigation and the Politics of Intangible Artifacts," *Technology and Culture* 55, no. 3 (2014): 623.

58. Longitude can be determined from the difference between the time at a known reference position (the Greenwich meridian, for instance), as registered by a faithful timepiece, and the local time, as deduced from celestial observations. Dava Sobel, *Longitude: The True Story of a Lone Genius Who Solved the Greatest Scientific Problem of His Time* (London: Harper Perennial, 1995).

59. Improvements in the precision and accuracy of atomic clocks are always being made. As of 2017, the most sophisticated atomic clock ever devised can keep time within an error of a second for fifteen billion years—longer than the age of the universe. In principle, such clocks could support satellite geopositioning at scales of under a millimeter.

60. Of equal importance to the calculation of distance is information on the exact orbital position (or *ephemeris*) of the satellite, which therefore is also included in the time-coded signal transmission from it.

61. Michael Russell Rip and David P. Lusch, "The Precision Revolution: The Navstar Global Positioning System in the Second Gulf War," *Intelligence and National Security* 9, no. 2 (1994): 178.

62. Michael Russell Rip and James M. Hasik, *The Precision Revolution: GPS and the Future of Aerial Warfare* (Annapolis, Md.: Naval Institute Press, 2002), 126–27.

63. Rip and Lusch, "Precision Revolution," 195.

64. Seth Schiesel, "On the Ground in Iraq, the Best Compass Is in the Sky," *New York Times,* April 17, 2003, http://www.nytimes.com/2003/04/17/technology/on-the-ground-in-iraq-the-best-compass-is-in-the-sky.html.

65. *Defense Science Board Task Force on the Future of the Global Positioning System* (Washington, D.C.: Office of the Under Secretary of Defense for Acquisition, Technology, and Logistics, October 2005), 33, http://www.acq.osd.mil/dsb/reports/ADA443573.pdf.

66. Positioning System for Deep Ocean Navigation (POSYDON), DARPA-BAA-15-30, DARPA Strategic Technology Office, April 23, 2015, 6.

67. Defense Advanced Research Projects Agency, Fiscal Year 2011 Budget Estimates, Department of Defense, February 2010, 33.

68. Defense Advanced Research Projects Agency, Fiscal Year 2011 Budget Estimates, Department of Defense, February 2010, 33. A further DARPA research project has investigated the potential for harnessing the very-low-frequency radio pulses emitted by lightning strikes (whether naturally occurring or artificially generated) into a system of underground geolocation that might support the subterranean naviga-

tion of troops. "Underground Mole-Satnavs to Work Off Lightning Strikes," *Register,* March 10, 2010, http://www.theregister.co.uk/2010/03/10/darpa_s_bug/.

69. Ryan Bishop, "Project 'Transparent Earth' and the Autoscopy of Aerial Targeting: The Visual Geopolitics of the Underground," *Theory, Culture, and Society* 28, no. 7–8 (2011): 272.

70. MacKenzie, *Inventing Accuracy.*

71. U.S. Patent 2,526,682 awarded to James E. Bellit and Henry C. Mulberger (October 24, 1950), 5.

72. Richard F. Koch and Donald C. Evans, "ATRAN Terrain Sensing Guidance—the Grand-Daddy System," *Image Processing for Missile Guidance: Proceedings of the Society of Photo-Optical Instrumentation Engineers* 238 (1980): 2–9.

73. In addition to inertial guidance and TERCOM, Tomahawk cruise missiles typically make use of GPS and digital scene matching for direction to their targets. The most accurate long-range weapons systems thus rely on a complementary suite of guidance technologies that correct and compensate for their individual imprecisions and vulnerabilities.

74. George M. Siouris, *Missile Guidance and Control Systems* (New York: Springer-Verlag, 2004), 551–55.

75. W. Seth Carus, *Cruise Missile Proliferation in the 1990s* (Westport, Conn.: Praeger, 1992), 55.

76. Jay L. Larson and George A. Pelletiere, *Earth Data and New Weapons* (Washington, D.C.: National Defense University, 1989).

77. Wilson Andrews and Todd Lindeman, "$52.6 billion: The Black Budget," *Washington Post,* August 29, 2013, http://www.washingtonpost.com/wp-srv/special/national/black-budget/.

78. James Gleick, *The Information: A History, a Theory, a Flood* (London: Fourth Estate, 2011).

79. The following analysis of the map communication model owes much to Jeremy Crampton's insightful treatment of "how mapping became scientific" in Crampton, *Mapping,* 49–61.

80. Arthur H. Robinson, "Geographic Cartography," in James and Jones, *American Geography,* 558.

81. Arthur H. Robinson, *The Look of Maps* (Madison: University of Wisconsin Press, 1952).

82. Arthur H. Robinson, Joel L. Morrison, and Phillip C. Muehrcke, "Cartography, 1950–2000," *Transactions of the Institute of British Geographers* 2, no. 1 (1977): 6.

83. See Bousquet, *Scientific Way of Warfare,* for a general discussion of the role of information theory and cybernetics in postwar military organization.

84. Claude E. Shannon, "A Mathematical Theory of Communication," *Bell System Technical Journal* 27, no. 3 (1948).

85. Joel L. Morrison, "The Science of Cartography and Its Essential Processes," *Cartographica* 4, no. 1 (1977): 69.

86. Anton Kolácný, "Cartographic Information: A Fundamental Concept and Term in Modern Cartography," *Cartographic Journal* 6, no. 1 (1969): 48.

87. Arthur H. Robinson and Barbara B. Petchenik, *The Nature of Maps* (Chicago: University of Chicago Press, 1976), 24.

88. Daniel R. Montello, "Cognitive Map-Design Research in the Twentieth Century: Theoretical and Empirical Approaches," *Cartography and Geographic Information Science* 29, no. 3 (2002): 283.

89. Arthur H. Robinson and Barbara B. Petchenik, "The Map as a Communication System," *Cartographica* 14, no. 1 (1977): 108.

90. Alan M. MacEachren, *How Maps Work: Representation, Visualization, and Design* (New York: Guilford Press, 1995), 7–8.

91. Robinson et al., "Cartography, 1950–2000," 6.

92. Christopher Board, "Cartographic Communication," *Cartographica* 18, no. 2 (1981): 46.

93. Pickles, *History of Spaces*; Crampton, *Mapping.*

94. Andrew Pickering, "Cyborg History and the World War II Regime," *Perspectives on Science* 3, no. 1 (1995).

95. Department of Environment, *Handling Geographic Information* (London: HSMO, 1987): 132.

96. Tobler was notably involved in the development of the SAGE (Semi-Automatic Ground Environment) air-defense system, for which he plotted computer-generated maps. Trevor J. Barnes, "Geography's Underworld: The Military-Industrial Complex, Mathematical Modelling and the Quantitative Revolution," *Geoforum* 39, no. 1 (2008): 11–12. For more details on SAGE, see Paul N. Edwards, *The Cold War: Computers and the Politics of Discourse in Cold War America* (Cambridge, Mass.: MIT Press, 1996).

97. Waldo R. Tobler, "Automation and Cartography," *Geographical Review* 49, no. 4 (1959): 526.

98. Martin Dodge, Rob Kitchin, and Chris Perkins, eds., *Rethinking Maps* (London: Routledge, 2009), 7. For MacEachren, geographic visualization (GVIS) is "the most important development in cartography since the thematic mapping 'revolution' of the early nineteenth century. For map users, GVIS represents nothing less than a new way to think spatially. For cartographers, GVIS represents a substantial change in emphasis from maps as a presentation tool to maps as part of a thinking-knowledge construction process." MacEachren, *How Maps Work,* 560.

99. Alan M. MacEachren and Menno-Jan Kraak, "Research Challenges in Geovisualization," *Cartography and Geographic Information Science* 28, no. 1 (2001): 3.

100. Pickles, *History of Spaces,* 156.

101. Barnes, "Geography's Underworld," 11; Waldo R. Tobler, "Analytical Cartography," *American Cartographer* 3, no. 1 (1976): 21. See also Keith C. Clarke and John G. Cloud, "On the Origins of Analytical Cartography," *Cartography and Geographic Information Science* 27, no. 3 (2000), and John Cloud, "American Cartographic Transformations during the Cold War," *Cartography and Geographic Information Science* 29, no. 3 (2002).

102. Oliver Belcher, "Data Anxieties: Objectivity and Difference in Early Vietnam War Computing," in *Algorithmic Life: Calculative Devices in the Digital Age,* ed. Louise Amoore and Volha Piotukh (London: Routledge, 2016).

103. Neil Smith, "History and Philosophy of Geography: Real Wars, Theory Wars," *Progress in Human Geography* 16, no. 2 (1992): 257.

104. Keith C. Clarke, "Maps and Mapping Technologies of the Persian Gulf War," *Cartography and Geographic Information Systems* 19, no. 2 (1992): 84.

105. Harry Greene, Larry Stotts, Ryan Paterson, and Janet Greenberg, "Command Post of the Future: Successful Transition of a Science and Technology Initiative to a Program of Record," *Defense Acquisition Review Journal* 17, no. 1 (2010): 6.

106. Kathryn Bailey, "CPOF's Capabilities Evolve into Web-Based Environment," *Fort Gordon Signal,* December 12, 2014, http://www.ftgordonsignal.com/news/2014-12-19/Community_Events/CPOFs_capabilities_evolve_into_webbased_environmen.html.

107. "The most crucial needs are often related to cognitive limitations and biases, not obvious to developers and rarely known to the operators themselves (and sometimes even denied by them)." Alexander Kott, "Concluding Thoughts," in Kott, *Battle of Cognition,* 217.

108. Mica R. Endsley, "Design and Evaluation for Situation Awareness Enhancement," *Proceedings of the Human Factors and Ergonomics Society Annual Meeting* 32, no. 2 (1988): 97.

109. Greene et al., "Command Post of the Future," 6.

110. Caroline Croser, "Networking Security in the Space of the City: Event-ful Battlespaces and the Contingency of the Encounter," *Theory and Event* 10, no. 2 (2007): § 36. See also Caroline Croser, *The New Spatiality of Security: Operational Uncertainty and the U.S. Military in Iraq* (London: Routledge, 2011). On CPOF and the wider practices of digital mapping in American counterinsurgency operations, see Derek Gregory, "Seeing Red: Baghdad and the Event-ful City," *Political Geography* 29, no. 5 (2010).

111. Caroline Croser, *Organising Complexity: Modes of Behaviour in a Networked Battlespace* (Duntroon, Australia: Land Warfare Studies Centre, 2007), 36.

112. Thrift, *Non-representational Theory,* 89.

113. Wolfgang Förstner, "Mapping on Demand—A Dream" (2002), 1, http://www.ipb.uni-bonn.de/uploads/tx_ikgpublication/foerstner02.mapping_on_demand_dream.pdf; Eric Coppock, Dennis Nicks Jr., and Roy Nelson, "Real-Time Creation and Dissemination of Digital Elevation Mapping Products Using Total Sight™ Flash Lidar," paper presented at the American Society for Photogrammetry and Remote Sensing (ASPRS) 2011 annual conference, Milwaukee, Wis., http://www.asprs.org/a/publications/proceedings/Milwaukee2011/files/Coppock.pdf; "Rapid 3D Mapping Gives You Superior Awareness," (Stockholm, Saab Group, 2011), https://web.archive.org/web/20120106015143/http://www.saabgroup.com/Global/Documents%20and%20Images/Civil%20Security/Police%20and%20Rescue%20Solutions/Rapid%203D%20Mapping/Rapid3D%2013.pdf; Maria Callier, "Portable Laser Backpack Revolutionizes 3D Mapping," Wright-Patterson Air Force Base, September 7, 2010, http://www.wpafb.af.mil/News/Article-Display/Article/400060/portable-laser-backpack-revolutionizes-3d-mapping/; Kristina Grifantini, "Making 3D Maps on the Move," *MIT Technology Review,* November 18, 2009, http://www.technologyreview.com/news/416331/making-3d-maps-on-the-move/.

114. Jorge Luis Borges, *Collected Fictions,* trans. Andrew Hurley (London: Penguin Books, 1999), 325.

115. "Present-day simulators try to make the real, all the real, coincide with their simulation models. But it is no longer a question of either maps or territory. Something has disappeared: the sovereign difference between them." Jean Baudrillard, *Simulations,* trans. Phil Beitchman, Paul Foss, and Paul Patton (New York: Semiotext(e), 1983), 2.

5. Hiding

1. Ralph Peters, "The Age of Fatal Visibility," *Military Review* 68, no. 8 (1988).
2. Libicki, "Information and Nuclear RMAs Compared," 2.
3. Jünger, *On Pain,* 39.
4. Deleuze and Guattari, *A Thousand Plateaus,* 277.
5. Julian Trevelyan, "The Technique of Camouflage," *Architectural Review* 96, no. 573 (1944): 68.

6. John L. Scott, László Moholy-Nagy, and György Kepes, "Civilian Camouflage Goes into Action," *Civilian Defense* 1, no. 2 (1942), 10.

7. National Research Council, *Psychology for the Fighting Man* (Washington, D.C.: Infantry Journal, 1943), 76.

8. Anthony King, "The Digital Revolution: Camouflage in the Twenty-First Century," *Millennium* 42, no. 2 (2014): 401.

9. J. F. C. Fuller, *The Conduct of War, 1789–1961: A Study of the Impact of the French, Industrial, and Russian Revolutions on War and Its Conduct* (London: Eyre & Spottiswoode, 1962), 140.

10. William Hodson, *Twelve Years of a Soldier's Life in India* (Boston, Mass.: Ticknor and Fields, 1860), 120.

11. Jünger, *On Pain,* 39.

12. Col. George Squier, U.S. military attaché to Great Britain (1915), quoted in Finnegan, *Shooting the Front,* 46.

13. James Philip Robinson, "'Darkened Surfaces': Camouflage and the Nocturnal Observation of Britain, 1941–45," *Environment and Planning A* 45, no. 5 (2013).

14. Roy R. Behrens, *Camoupedia: A Compendium of Research on Art, Architecture and Camouflage* (Dysart, Iowa: Bobolink Books, 2009), 20.

15. Peter Forbes, *Dazzled and Deceived: Mimicry and Camouflage* (New Haven, Conn.: Yale University Press, 2009), 151.

16. Hanna Rose Shell, *Hide and Seek: Camouflage, Photography, and the Media of Reconnaissance* (New York: Zone Books, 2012), 19.

17. Isla McLean Forsyth, "From Dazzle to the Desert: A Cultural-Historical Geography of Camouflage" (Ph.D. diss., University of Glasgow, 2012), 205–8.

18. Behrens, *Camoupedia,* 94.

19. Before World War I, Thayer's negotiations with the U.S. Navy over his proposed ship camouflage schemes bore no fruit, although he did obtain a 1902 patent for a "Process of Treating the Outsides of Ships, etc. for Making Them Less Visible." His attempted lobbying of the British War Office during the Great War was no more successful, but he did exert an indirect influence through the service of students and supporters in the nascent camouflage units.

20. Abbott H. Thayer, "The Law which Underlies Protective Coloration," *Auk* 13, no. 2 (1896): 127.

21. Roy R. Behrens, "The Theories of Abbott H. Thayer: Father of Camouflage," *Leonardo* 21, no. 3 (1988): 293.

22. Gerald H. Thayer and Abbott H. Thayer, *Concealing Coloration in the Animal Kingdom: An Exposition of the Laws of Disguise through Color and Pattern* (New York: Macmillan, 1909), 77.

23. Lucien-Victor Guirand de Scévola, "Souvenirs du Camouflage (1914–1918)," *Revue des deux Mondes* (December 1949): 720. Among the more renowned cubists to have served in French camouflage units during the war were André Mare and Jacques Villon (Marcel Duchamp's older brother). Picasso is reported to have exclaimed, on spotting a camouflaged artillery piece in the streets of Paris, "It's we who have done that!" For all this, one should not overstate the influence of cubism in the development of military camouflage, given the prominent roles also played by more conventional artists such as Abbott Thayer, Norman Wilkinson, and Solomon J. Solomon.

24. Cécile Coutin, *Tromper l'ennemi: L'invention du camouflage moderne en 1914–1918* (Paris: Editions Pierre de Taillac et Ministère de la Défense, 2012), 57.

25. For an account of German efforts, see Christoph Nübel, "Modern Warfare:

Camouflage Tactics ('Tarnung') in the German Army during the First World War," *First World War Studies* 6, no. 2 (2015).

26. Col. G. H. Addison, *Work of the Royal Engineers in the European War, 1914–1918: Camouflage Service* (Uckfield, U.K.: Naval & Military Press, 2006), 149.

27. Roy M. Stanley, *To Fool a Glass Eye: Camouflage versus Photoreconnaissance in World War II* (Shrewsbury, U.K.: Airlife, 1998), 67–70.

28. Behrens, "Theories of Abbott H. Thayer," 293.

29. As British zoologist Hugh B. Cott later observed, the term "dazzle" is something of a misnomer in that the term's customary usage evokes a momentary blinding by bright lights. The original American slang of "razzle-dazzle," from which the term is derived, or the alternative name of "baffle painting" are accordingly both more appropriate labels for the technique under consideration, but because the designation "dazzle" has endured, it will be privileged here too. However, it is important to note that the principles governing dazzle camouflage are distinct from those constitutive of the approach that we will examine under the term "bedazzlement." Hugh B. Cott, *Adaptive Coloration in Animals* (London: Methuen, 1940), 47.

30. J. Graham Kerr, "'Camouflage' of Ships in War," *Nature* 103, no. 2585 (May 15, 1919): 204.

31. Kerr, "Camouflage," 204.

32. Norman Wilkinson, "British Naval Camouflage" (1922), in *Ship Shape: A Dazzle Camouflage Sourcebook,* ed. Roy R. Behrens (Dysart, Iowa: Bobolink Books, 2012), 47.

33. Guy Hartcup, *Camouflage: A History of Concealment and Deception in War* (Barnsley, U.K.: Pen & Sword Books, 2008), 45–46.

34. See, for example, the joint issue of Lindell Theodore Bates, *The Science of Low Visibility and Deception against Attack by Submarines,* and Lloyd Ancile Jones, *Protective Coloration as a Means of Defense against Submarines* (New York: Submarine Defense Association, 1918).

35. "A Theatre for Studying Camouflaged Ship Models," *Scientific American* 88, no. 2290 (December 13, 1919): 348.

36. James Philip Robinson, "Invisible Targets, Strengthened Morale: Static Camouflage as a 'Weapon of the Weak,'" *Space and Polity* 16, no. 3 (2012).

37. Wolfgang Metzger, *Laws of Seeing,* trans. Lothar Spillmann (1936; Cambridge, Mass.: MIT Press, 2009), 85.

38. National Defense Research Committee, *Visibility Studies and Some Applications in the Field of Camouflage,* Summary Technical Report of Division 16 (Washington, D.C.: National Defense Research Committee, 1946), ix.

39. Thomas O'Neill and William L. Johnsmeyer, *DUAL-TEX: Evaluation of Dual-Texture Gradient Pattern* (West Point, N.Y.: Office of Military Leadership, U.S. Military Academy, 1977).

40. Alvin O. Ramsley and William B. Bushnell, *Development of the U.S. Woodland Battle Dress Uniform* (Natick, Mass.: United States Army Natick Research and Development Laboratories, 1981).

41. On fractals and nonlinearity more generally, see Bousquet, *Scientific Way of Warfare,* 163–83.

42. Conroy Nachtigall, "Disappear Us: Algorithms, Aesthetics, and the Armed Forces," *032c* 23 (Winter 2012–13).

43. U.S. patent US6805957 B1, "Camouflage U.S. Marine Corps Utility Uniform: Pattern, Fabric, and Design" (October 19, 2004).

44. Vincent A. Billock, Douglas W. Cunningham, and Brian H. Tsou, "What Visual

Discrimination of Fractal Textures Can Tell Us about Discrimination of Camouflaged Targets," paper presented at the Human Factors Issues in Combat Identification Workshop, Gold Canyon, Ariz., May 13, 2008, 3–4.

45. Silvia Mollicchi, "Flatness versus Depth: A Study of Algorithmically Generated Camouflage," *Security Dialogue* 48, no. 1 (2017).

46. Precursors of digital camouflage can be found in the German experiments with multiscale patterns in the Nazi era. These were subsequently printed on Waffen-S.S. uniforms during the war, later bequeathing the mottled "Flecktarn" pattern still in use in today's Bundeswehr and copied by numerous national armed forces. The underlying visual principles are essentially the same as those of digital camouflage, with the key difference lying in the latter's shift to a computational process for its design.

47. Matthew Cox, "Army Unveils Design Changes for New Camo Uniform," Military.com, August 6, 2014, http://www.military.com/daily-news/2014/08/06/army-unveils-design-changes-for-new-camo-uniform.html.

48. Interservice rivalry seems to have also played a role, with the Army apparently determined to outdo the Marines' own MARPAT scheme and more concerned with doing so on aesthetic appearance than on functional efficacy. See King, "Digital Revolution."

49. Billock, Cunningham, and Tsou, "What Visual Discrimination of Fractal Textures Can Tell Us," 11.

50. Department of the Army, *United States Army Field Manual FM 20-3: Camouflage, Concealment, and Decoys* (Washington, D.C.: Department of the Army, November 2010), 3.11.

51. The radar cross-sectional specification of a given aircraft can only ever be a rough estimate because even on stealth planes it will vary considerably, depending on the angle at which radar signals strike it.

52. Department of the Army, *United States Army Field Manual FM 20-3,* 3.3.

53. Carlo Kopp, "Russian VHF Counter Stealth Radars Proliferate," *Defence Today* 7, no. 3 (2008).

54. Westra, "Radar versus Stealth."

55. U.S. Commission on National Security/21st Century, *New World Coming: American Security in the Twenty-first Century* (Washington, D.C.: U.S. Government Printing Office, 1999), 49.

56. D. Schurig, J. J. Mock, B. J. Justice, S. A. Cummer, J. B. Pendry, A. F. Starr, and D. R. Smith, "Metamaterial Electromagnetic Cloak at Microwave Frequencies," *Science* 314, no. 5801 (2006); Jason Valentine, Jensen Li, Thomas Zentgraf, Guy Bartal, and Xiang Zhang, "An Optical Cloak Made of Dielectrics," *Nature Materials* 8, no. 7 (2009).

57. Investigations into the possibility of sound cloaking are also being undertaken presently. Lucian Zigoneanu, Bogdan-Ioan Popa, and Steven A. Cummer, "Three-Dimensional Broadband Omnidirectional Acoustic Ground Cloak," *Nature Materials* 13, no. 4 (2014).

58. See, for example, the following paper presenting "the design, mathematical modelling and analysis of a dynamic biomimetic pattern generation system using bio-inspired artificial chromatophores." Aaron Fishman, Jonathan Rossiter, and Martin Homer, "Hiding the Squid: Patterns in Artificial Cephalopod Skin," *Journal of the Royal Society Interface* 12, no. 108 (2015): 1.

59. Masahiko Inami, Naoki Kawakami, and Susumu Tachi, "Optical Camouflage Using Retro-Reflective Projection Technology," in *Proceedings of the Second IEEE and ACM International Symposium on Mixed and Augmented Reality* (2003), http://tachilab.org/content/files/publication/ic/inami2003ISMAR.pdf.

60. National Defense Research Committee, *Visibility Studies,* 225–41.

61. BAE Systems, "ADAPTIV—Cloak of Invisibility," http://www.baesystems.com/en-uk/feature/adativ-cloak-of-invisibility.

62. Robert Dieudonné, "Camoufleurs," *La Baïonnette* 112 (August 23, 1917): 534.

63. Xavier Boissel, *Paris est un leurre: La veritable histoire du faux Paris* (Paris: Éditions Inculte, 2012).

64. Seymour Reit, *Masquerade: The Amazing Camouflage Deceptions of World War II* (New York: Hawthorn, 1978), 61.

65. Hillel Schwartz, *The Culture of the Copy: Striking Likenesses, Unreasonable Facsimiles* (New York: Zone Books, 2014), 166.

66. Reit, *Masquerade,* 189.

67. Stanley, *To Fool a Glass Eye,* 78–80.

68. Isla Forsyth, "Designs on the Desert: Camouflage, Deception and the Militarization of Space," *Cultural Geographies* 21, no. 2 (2014).

69. Rick Beyer and Elizabeth Sayes, *The Ghost Army of World War II: How One Top-Secret Unit Deceived the Enemy with Inflatable Tanks, Sound Effects, and Other Audacious Fakery* (New York: Princeton Architectural Press, 2015).

70. Reit, *Masquerade,* 11–13.

71. David M. Glantz, *Soviet Military Deception in the Second World War* (London: Routledge, 1989).

72. Stanley, *To Fool a Glass Eye,* 71.

73. Kittler, *Optical Media,* 173.

74. Countermortar and counterbattery radar can be similarly overloaded if sufficient rounds are fired simultaneously.

75. *Defense Science Board Task Force on the Future of the Global Positioning System.*

76. Sonar decoys rely on the same principle as chaff or flares in frustrating sound-ranging apparatuses in the aquatic domain.

77. Martin C. Libicki, *The Mesh and the Net: Speculations on Armed Conflict in a Time of Free Silicon* (Washington, D.C.: National Defense University, 1994), 23.

78. Ralph Peters, "The Future of Armored Warfare," *Parameters* 27, no. 3 (1997): 53.

79. Gastón Gordillo, "Terrain as Insurgent Weapon: An Affective Geometry of Warfare in the Mountains of Afghanistan" *Political Geography* 64 (2018).

80. Stephen Graham, *Cities under Siege: The New Military Urbanism* (London: Verso Books, 2011), 160.

81. David Kilcullen, *Out of the Mountains: The Coming Age of the Urban Guerrilla* (London: Hurst, 2013).

82. Arthur Herman, "Notes from the Underground: The Long History of Tunnel Warfare," *Foreign Policy,* August 26, 2014.

83. Paul Virilio, *Bunker Archaeology,* trans. George Collins (New York: Princeton Architectural Press, 1994), 44.

84. Eric M. Sepp, "Deeply Buried Facilities: Implications for Military Operations," Occasional Paper No. 14 (Montgomery, Ala.: Center for Strategy and Technology, Air War College, Maxwell Air Force Base, 2000).

85. Libicki, *The Mesh and the Net*; John Arquilla and David Ronfeldt, *Swarming and the Future of Conflict* (Santa Monica, Calif.: RAND Corporation, 2000); David S. Alberts and Richard E. Hayes, *Power to the Edge: Command and Control in the Information Age* (Washington, D.C.: Department of Defense Command and Control Research Program, 2003).

86. Robert O. Work and Shawn Brimley, *20YY: Preparing for War in the Robotic Age* (Washington, D.C.: Center for a New American Security, 2014), 28–29.

87. Sean J. A. Edwards, *Swarming on the Battlefield: Past, Present, and Future* (Santa Monica, Calif.: RAND Corporation, 2000).

88. For further discussion of the tactic of swarming and the corresponding model of "chaoplexic warfare," see Bousquet, *Scientific Way of Warfare,* 185–233. For a critical reading of the modern discourse of swarming, see also Daniel Bertrand Monk, "Hives and Swarms: On the 'Nature' of Neoliberalism and the Rise of the Ecological Insurgent," in *Evil Paradises: Dreamwolds of Neoliberalism,* ed. Mike Davis and Daniel Bertrand Monk (New York: New Press, 2008), 262–73.

89. Paul Sharre, *Robotics on the Battlefield, Part 2: The Coming Swarm* (Washington, D.C.: Center for a New American Security, 2014), 20.

90. Achille Mbembe, "Necropolitics," *Public Culture* 15, no. 1 (2003): 36.

91. Reza Negarestani, "The Militarization of Peace: Absence of Terror or Terror of Absence?" in *Collapse,* ed. Robin Mackay (Oxford: Urbanomic, 2007), 1:56–58.

92. Negarestani, "Militarization of Peace," 1:90.

93. A guidebook released in 2015 by the Islamic State of Iraq and Syria (ISIS) for the attention of "lone wolf Mujahideen and small cells" explains that "the U.S. today use technology to fight us, they use drones, spies among the Mujahideen. It's a new form of asymmetric warfare. They don't know us, they don't see us, until they get enough info to reach us. So our war today is a covert war, and to ensure victory, you have to be cautious." The manual goes on to recommend that any jihadist working on a "secret operation" should avoid displaying any visible attachment to religion such as wearing traditional Islamic garb, sporting a beard, or regularly frequenting a mosque. Abu Ubayda Abdullah al-Adm, "Safety and Security Guidelines for Lone Wolf Mujahideen and Small Cells" (n.p.: Al-Fajr Media Center, 2015).

94. Antoine Bousquet, "The Battlefield is Dead," Aeon, October 9, 2017, https://aeon.co/essays/how-the-bloody-field-of-battle-made-way-for-precision-drones. See also Mégret, "War and the Vanishing Battlefield."

95. Xavier Guillaume, Rune S. Andersen and Juha A. Vuori, "Paint It Black: Colours and the Social Meaning of the Battlefield," *European Journal of International Relations* 22, no. 1 (2016).

96. Toni Pfanner, "Military Uniforms and the Laws of War," *International Review of the Red Cross* 86, no. 853 (2004).

97. For a nonexhaustive catalog of such instances from 1904 to 1991, see W. Hay Parks, "Special Forces' Wear of Non-standard Uniforms," *Chicago Journal of International Law* 4, no. 2 (2003).

98. Barbara Starr, Chris Lawrence, and Joe Sterling, "ISAF: Insurgents in Deadly Attack in Afghanistan Wore U.S. Army Uniforms," CNN, September 16, 2012, http://edition.cnn.com/2012/09/14/world/asia/afghanistan-fatal-attack/; "Gunman in Afghan Uniform Kills Two U.S. Troops at Base," Al Jazeera, August 26, 2015, http://america.aljazeera.com/articles/2015/8/26/gunman-in-afghan-uniform-kills-2-us-troops-at-base.html.

99. Charles L. Smith, "Soviet Maskirovka," *Airpower Journal* 2, no. 1 (Spring 1988), http://www.airpower.maxwell.af.mil/airchronicles/apj/apj88/spr88/smith.html.

100. Brian Massumi, *Ontopower: War, Powers, and the State of Perception* (Durham, N.C.: Duke University Press, 2015).

101. Ian Shaw and Majed Akhter, "The Dronification of State Violence," *Critical Asian Studies* 46, no. 2 (2014).

102. Chamayou, *Theory of the Drone,* 34; Steven Marks, Thomas Meer, and Matthew Nilson, "Manhunting: A Methodology for Finding Persons of National Interest" (mas-

ter's thesis, Naval Postgraduate School, Monterey, Calif., 2005), 13, http://www.dtic.mil/docs/citations/ADA435585.

Conclusion

1. Voelz, "Individualization of American Warfare."

2. Bousquet, "War."

3. Frédéric Gros, *States of Violence: An Essay on the End of War* (London: Seagull Books, 2010), 262.

4. Gregory, "Everywhere War," 238.

5. Scott Wilson and Al Kamen, "'Global War on Terror' Is Given New Name," *Washington Post,* March 25, 2009, http://www.washingtonpost.com/wp-dyn/content/article/2009/03/24/AR2009032402818.html.

6. Julian Reid, *The Biopolitics of the War on Terror: Life Struggles, Liberal Modernity and the Defence of Logistical Societies* (Manchester: Manchester University Press, 2006); Claudia Aradau and Rens van Munster, "Governing Terrorism through Risk: Taking Precautions, (Un)Knowing the Future," *European Journal of International Relations* 13, no. 1 (2007); Louise Amoore, *The Politics of Possibility: Risk and Security beyond Probability* (Durham, N.C.: Duke University Press, 2013).

7. Lucy Suchman, Karolina Follis, and Jutta Weber, "Tracking and Targeting: Sociotechnologies of (In)security," *Science, Technology, and Human Values* 42, no. 6 (2017).

8. Peter B. Kraska, "Militarization and Policing—Its Relevance to 21st Century Police," *Policing* 1, no. 4 (2007); Radley Balko, *Rise of the Warrior Cop: The Militarization of America's Police Forces* (New York: Public Affairs, 2013).

9. Terry Frieden, "Holder Does Not Rule Out Drone Strike Scenario in U.S.," CNN, March 6, 2013, http://edition.cnn.com/2013/03/05/politics/obama-drones-cia/.

10. Rens van Munster, "The War on Terrorism: When the Exception Becomes the Rule," *International Journal for the Semiotics of Law* 17, no. 2 (2004); Jef Huysmans, "The Jargon of Exception—on Schmitt, Agamben and the Absence of Political Society," *International Political Sociology* 2, no. 2 (2008); Louise Amoore and Marieke de Goede, eds., *Risk and the War on Terror* (New York: Routledge, 2008).

11. David A. Deptula, "Evolving Technologies and Warfare in the 21st Century: Introducing the 'Combat Cloud,'" *Mitchell Institute Policy Papers* 4 (September 2016).

12. David A. Deptula, "'Combat Cloud' is 'New Face of Long-Range Strike,'" *Armed Forces Journal,* September 18, 2013, http://armedforcesjournal.com/deptula-combat-cloud-is-new-face-of-long-range-strike/.

13. Paul Virilio, *Desert Screen: War at the Speed of Light,* trans. Michael Degener (London: Continuum, 2005), 4. Virilio writes of a "transpolitical" epoch in which the evacuation of meaningful politics follows from the autonomization of the war machine. "The 'automatic' character, no longer solely of arms, of means, but also and above all of command, amounts to the denial of the capacities of reasoning and therefore, finally of all 'politics' to the advantage of systems of information and 'transpolitical' decision." Paul Virilio, *Negative Horizon,* trans. Michael Degener (New York: Continnum, 2006), 181.

14. Deptula, "Introducing the 'Combat Cloud,'" 5.

15. Deptula, "Introducing the 'Combat Cloud,'" 9.

16. Bradley Peniston, "Army Warns that Future War with Russia or China Would Be 'Extremely Lethal and Fast,'" Defense One, October 18, 2016, http://www.defenseone.com/threats/2016/10/future-army/132105/.

17. Martin Heidegger, "The Age of the World Picture" (1938), in *Martin Heidegger: Off the Beaten Track,* ed. and trans. Julian Young and Kenneth Haynes (Cambridge: Cambridge University Press, 2002), 67.

18. Martin Heidegger, *Nietzsche, Volume 4: Nihilism,* trans. Frank A. Capuzzi (San Francisco, Calif.: Harper & Row, 1991), 28.

19. Rey Chow, *The Age of the World Target: Self-Referentiality in War, Theory, and Comparative Work* (Durham, N.C.: Duke University Press, 2006), 12.

20. Hito Steyerl, "In Free Fall: A Thought Experiment on Vertical Perspective," *e-flux journal* 24 (April 2011), http://www.e-flux.com/journal/in-free-fall-a-thought-experiment-on-vertical-perspective/.

21. Martin Heidegger, "Why Poets?," in *Martin Heidegger: Off the Beaten Track,* 217.

22. Paul Scharre, "Centaur Warfighting: The False Choice of Humans vs. Automation," *Temple International and Comparative Law Journal* 30, no. 1 (2016): 151.

23. Astrid H. M. Nordin and Dan Öberg, "Targeting the Ontology of War: From Clausewitz to Baudrillard," *Millennium* 43, no. 2 (2015): 400.

24. Major F. A. de V. Robertson, "The Air Exercises," *Flight* 23, no. 31 (July 31, 1931): 764.

25. The aircraft's altitude necessary for the calculation was obtained either by prearrangement, visual surveying, or wireless communication from the pilot.

INDEX

Page numbers in italic refer to illustrations.

ANTOINE BOUSQUET is reader in international relations at Birkbeck, University of London. He is author of *The Scientific Way of Warfare: Order and Chaos on the Battlefields of Modernity.*